Fresh Air

KEVIN,

THANKS FOR MAKING US A PART OF YOUR TEAM!

— Mark

Fresh Air

Marketing Gurus on Radio

The Best Interviews from the Radio Marketing Nexus

edited by Mark Ramsey

iUniverse, Inc.
New York Lincoln Shanghai

Fresh Air
Marketing Gurus on Radio

Copyright © 2005 by Mark R. Ramsey

All rights reserved. No part of this book may be used or reproduced by any means, graphic, electronic, or mechanical, including photocopying, recording, taping or by any information storage retrieval system without the written permission of the publisher except in the case of brief quotations embodied in critical articles and reviews.

iUniverse books may be ordered through booksellers or by contacting:

iUniverse
2021 Pine Lake Road, Suite 100
Lincoln, NE 68512
www.iuniverse.com
1-800-Authors (1-800-288-4677)

ISBN-13: 978-0-595-37658-2 (pbk)
ISBN-13: 978-0-595-82044-3 (ebk)
ISBN-10: 0-595-37658-4 (pbk)
ISBN-10: 0-595-82044-1 (ebk)

Printed in the United States of America

For my wife, Jeanne, the source of all my fresh air

For more thoughtful marketing wisdom for the radio industry, please visit:

http://www.radiomarketingnexus.com

The Radio Marketing Nexus

Contents

Introduction 1
The Gurus 5
Content is King 14
Differentiate or Die 23
Spread the Buzz 25
Tell a Good Story 28
Think "Creativity" 31
How to be Funny 34
Market Across Generations 37
Energize your Brand Evangelists 40
The Radio Marketing Playbook 45
Become a Cult Brand 48
Sell the Dream 51
Stand for Something 54
Don't Be Average 58
Impact is Cheap 60
What's in a Name? 63
Create Evangelists 72
Better Ads, Happier Listeners 76

Inspiring Passion 82
The Future of Radio 85

Introduction

"Resolve to be always beginning"
—*Rainer Maria Rilke*

In the beginning, there was the magic.

The miracle of sound traveling invisibly through the air to that mysterious box with the dial. Wherever you go, it's with you, never more than the flip of a switch away.

For much of the audience, the listening remains (mostly) but the romance of radio has passed. Once the center of the home's entertainment universe, the source of the best information and the most entertainment, radio today is one of many media options and rarely the most exciting one. In the past few years alone the number of alternatives to faithful old radio has exploded. Today magical, invisible audio can meet your ears via the Internet, over your cell phone, via Satellite, through your mp3 player, and so on.

For much of its history radio has owned the only route to your ears. That kind of responsibility is a freedom and a curse. A freedom because it allows the creation of audio wonders great and small. A curse because the absence of competition from alternate technologies has a way of promoting a fat and happy settling, a bloated behemoth of an industry which knows how to do what it has always done, but not what it needs to do next. The status quo perpetuates nothing so much as more status quo.

In the years to come radio will experience the most profound challenges to its status quo ever. We are entering a time where smart thinking, novel ideas, and profound innovation will spell the difference between relevance for future generations and quaint nostalgic obsolescence.

Rising to these challenges will require break-out thinking.

Like so many mature industries, radio needs to be curious. We must not read the same trades, hire the same consultants, do the same research, enlist the same programmers, subscribe to the same assumptions, repurpose the same production, syndicate and voice-track the same personalities, and copy the same names, logos, and ideas until, in some markets, the only thing "local" about radio is the address of the sales office.

But where are these new ideas to come from?

That's what outsiders are for, to jar us from business as usual, to boldly proclaim that the emperor has no clothes. To shake us silly. I'm talking about that unique breed of outsiders, the marketing gurus whose ideas span many industries and shoot straight to the heart of what's next.

So here's my request: Pretend that everything you know is wrong.

I know, that's not likely. But stay with me here.

Suppose all the rules you operate by, all the "givens" about what listeners seem to want and how radio is supposed to be done, suppose all that doesn't exist.

Collected here are many of the best interviews conducted for the Radio Marketing Nexus (http://www.radiomarketingnexus.com), the notorious, controversial, and spin-free blog of Mercury Radio Research (http://www.mercradio.com).

This book marks the first time an all-star marketing "dream team" has assembled to take aim at the needs of the radio industry. Almost all of these experts have profound marketing and branding expertise with some of the best known and most respected corporate names around, and virtually all of them have little or no experience in the radio trenches.

That means some of their recommendations will strike you as naive. And maybe they are. But these folks are not just marketers, they're also radio listeners. So suspend your sense of "what works" and read their words with fresh eyes.

Because without fresh eyes, there can be no fresh air.

The Gurus

"I want to stay as close to the edge as I can without going over. Out on the edge you see all kinds of things you can't see from the center."
—Kurt Vonnegut

Tom Asacker

Tom Asacker is the author of *A Clear Eye for Branding* and *Sandbox Wisdom: Revolutionize your Brand with the Genius of Childhood*. He is a catalyst and a non-conformist, a popular lecturer, writer and corporate advisor with a strong background in business and brand strategy, as well as internal culture and operations development. He gets paid to provoke people to think about things in ways that are unclouded by the issues they deal with on a daily basis.

Asacker is a recipient of the George Land Innovator of the Year Award; he holds medical patents and product design awards; and he is recognized by *Inc.* magazine, MIT, and the Young Entrepreneurs' Organization as a past member of their "Birthing of Giants" entrepreneurial executive leadership program.

Dave Balter

Dave Balter is the co-author of *Grapevine: The New Art of Word-of-Mouth Marketing*. He is the Founder and President of BzzAgent, an innovative word-of-mouth marketing agency. He is also a founding member of the Word-of-Mouth Marketing Association (WOMMA), co-chair of WOMMA's Ethics Council, and a frequent speaker to corporations and universities on the subject of Word-of-Mouth and non-traditional media.

Dave was named to the "40 under 40" by the Advertising Specialty Institute in 2001. He holds a provisional patent in the process and systems of Word-of-Mouth marketing and research.

Scott Bedbury

"Scott Bedbury may be the greatest brand maven of our time," said Tom Peters. Bedbury is author of *A New Brand World: 8 Principles for Achieving Brand Leadership in the 21st Century* and CEO of Brandstream, Inc, an independent brand development consultancy with major clients worldwide.

In 1987, Bedbury left the advertising agency industry to direct Nike's worldwide marketing division. In seven years, he helped grow Nike from $750 million to $5 billion in revenues, while building one of the world's most successful campaigns, 'Just do it'. In 1994, Bedbury joined Starbucks as chief marketing officer. He helped redesign its stores, entered its products into dozens of new markets, developed Starbucks ice cream and bottled Frappuccino for the grocery channel and established global distribution partnerships such as United Airlines. His brand development strategies helped grow Starbucks from 350 stores to several thousand.

B. J. Bueno

Bolivar J. Bueno is an independent marketing consultant and cult branding expert. He is a lifelong student of marketing and the study of the powerful relationships between brands and customer loyalty. B. J. is the co-author of *The Power of Cult Branding*.

Seth Godin

Seth Godin is a bestselling author, entrepreneur and agent of change. Godin is author of six books that have been bestsellers around the world and changed the way people think about marketing, change and work. His best-known books include bestsellers *Permission Marketing*, *Unleashing the Ideavirus*, and *Purple Cow*. Seth is a renowned speaker as well. He was recently chosen as one of 21 Speakers for the Next Century by *Successful Meetings* and is consistently rated among the very best speakers by the audiences he addresses.

He holds an MBA from Stanford, and was called "the Ultimate Entrepreneur for the Information Age" by *Business Week*.

Andy Goodman

After founding the American Comedy Network, a radio syndication company, Andy Goodman spent three seasons writing and co-producing the ABC-TV show "Dinosaurs" (plus co-writing the pilot episode of "The Nanny"). Now Goodman heads *A Goodman*, a communications consultancy based in Los Angeles. He specializes in helping public interest groups, foundations, and progressive businesses communicate more effectively thought storytelling.

Doug Hall

Doug Hall is the author of *Jump Start Your Business Brain* and *Jump Start Your Marketing Brain*.

You may not know Doug Hall, but the odds are you know of his work. A national survey found that the average home utilizes eighteen products or services Doug and his Eureka! Inventing team have had a hand in.

A former "Master Marketing Inventor" at Procter & Gamble, Doug is the founder of Eureka! Inventing, a corporate think tank dedicated to turning the art of innovation into a reliable science. Clients include some of the biggest corporations in the world including American Express, The Ford Motor Company, Nike and Walt Disney.

Doug has been named by *Inc. Magazine*, *The Wall Street Journal*, A&E top 10 and Dateline NBC as "One of America's Top Innovation Experts." He is host of the public radio program *Brain Brew Idea Factory*.

Mark Hughes

Mark Hughes is the author of business bestseller *Buzzmarketing* and founder of Buzzmarketing alternative media. Previously, Mark was vice president of marketing for eBay's Half.com (the #1 Internet retailer of used books, music, and movies). Hughes grew eBay's Half.com from zero to 8 million registered users in less than three years.

Hughes marketed the start-up and literally put Half.com on the map by convincing Halfway, Oregon to rename itself to Half.com, Oregon…creating the first dot-com city in America.

The buzz generating event was dubbed by *Time* magazine as "one of the greatest publicity coups" in history.

Jay Jurisich

Jay Jurisich is the co-founder of San Francisco-based Igor, a maverick naming and branding agency focused on bringing myth and meaning back into company and product names. Clients include Canon, Palm, Nike, Gap, MTV, Amway, Hasbro, and many more.

Guy Kawasaki

Guy Kawasaki is a managing director of Garage Technology Ventures, an early-stage venture capital firm, and a columnist for Forbes.com. Previously, he was an Apple Fellow at Apple Computer, Inc. where he was one of the individuals responsible for the success of the Macintosh computer.

Guy is the author of eight books including *The Art of the Start, Rules for Revolutionaries, How to Drive Your Competition Crazy, Selling the Dream*, and *The Macintosh Way*. He has a BA from Stanford University and an MBA from UCLA as well as an honorary doctorate from Babson College.

Richard Laermer

A frequently quoted authority on media culture and hype, Richard Laermer is a former magazine and newspaper journalist and the author of five best-selling books.

Richard is head of RLM Public Relations, which specializes in building aggressive buzz and concrete brand images for blue-chip clients, including Fujifilm, Barnes & Noble Inc., Time-Life, and Allergan Inc.

He is author of the highly acclaimed, award-winning *Full Frontal PR: Getting People Talking About You and Your Business or Your Product.*

He was one of the stars of the TLC network reality program *Taking Care of Business.* His segment "Unspun" is a regular feature on NY1, and his columns on the PR business appear in *Advertising Age, Adweek* and *PR News.*

Anne Libera

Anne Libera is an Executive Artistic Director of Chicago's legendary Second City improvisational troupe, home to such comedy luminaries as John Belushi, Gilda Radner, Bill Murray, Tina Fey, Bonnie Hunt, and many more.

The author of the invaluable *The Second City Almanac of Improvisation,* Anne has reviewed theater on WGN radio and written for the NPR news quiz show *Wait, Wait, Don't Tell Me.* Anne serves on the governing board for Gilda's Club Chicago.

Ben McConnell

Ben McConnell is co-author (with Jackie Huba) of *Creating Customer Evangelists,* called the "new mantra for entrepreneurial success" by the *New York Times* and "an absorbing read" by Harvard Business School.

With Jackie, Ben pens regular columns for MarketingProfs.com and speaks frequently at industry, association and company conferences. They also facilitate the creation of customer evangelism plans inside organizations.

John Moore

For the past decade, John Moore has made his mark in the marketing world by creating, championing, and implementing marketing ideas and branding ideals for Whole Foods Market and Starbucks Coffee.

As the Director of National Marketing for Whole Foods Market, John focused his team on engaging in activities which were less about using traditional advertising and more about using the influential power of customers as the advertising vehicle. At Starbucks, John led countless highly successful in-store and out-of-store marketing promotions as a Retail Marketing Manager.

Through his Brand Autopsy Marketing Practice, John shares business and marketing advice with small-to-midsize companies aspiring to become the next Whole Foods or Starbucks. He is also the author of the Brand Autopsy blog (http://www.brandautopsy.com) and *Tribal Knowledge*, a business management book due in bookstores fall 2006.

Jack Trout

Jack Trout is president of Trout & Partners, one of the most prestigious marketing firms with headquarters in Greenwich, Connecticut, USA and offices in 13 countries. Jack manages and supervises a global network of experts that apply his concepts and develop his methodology around the world. The firm has worked for AT&T, IBM, Burger King, Merrill Lynch,

Xerox, Merck, Ericsson, KPMG Peat Marwick, Procter & Gamble, Southwest Airlines and other Fortune 500 companies.

With Al Ries he co-authored the business bookshelf classics, *Positioning: The Battle for Your Mind* and *The 22 Immutable Laws of Marketing*.

David Wolfe

David Wolfe is the author of *Ageless Marketing* and an internationally recognized consumer behavior consultant. He heads the Center for Ageless Marketing, where the prime directive is to compensate for shrinking young adult markets by using marketing tactics to get more sales from every adult age group—marketing based on values that appeal across generational divides.

David's client roster includes the likes of American Express, AT&T, Bausch & Lomb, Blue Cross of California, AARP, General Motors, and the Coca Cola Company.

Willy Wonka

Willy Wonka is rumored to be a fictional character but somehow materialized long enough for an interview.

Wonka is the owner of a legendary chocolate factory that creates the tastiest, most splendiferous chocolates known to man. Although he has not written a book, he is featured prominently in Roald Dahl's *Charlie and the Chocolate Factory* and two movies based on the book.

John Zagula

John Zagula is a founder of Ignition Partners, a Silicon Valley venture capital firm. He is also the co-author (along with Rich Tong) of the bestselling, acclaimed book, *The Marketing Playbook*.

Prior to Ignition, John spent over 8 years at Microsoft in various senior marketing and strategy positions. As GM of Global Campaigns he was responsible for Microsoft's overall business-to-business communications across all business targets, products and offerings. In previous positions he was responsible for the development of the Microsoft Office brand and marketing the desktop and server applications product lines. He contributed to the original competitive wins of Word and Excel and the re-positioning of desktop applications into Office.

Content is King

Courage is being scared to death—but saddling up anyway
—John Wayne

Doug Hall is an inventor, author, and revolutionary. He's the founder of the Eureka Ranch, a think tank and idea factory. His clients include Ford, Disney, Procter & Gamble, Nike, Johnson & Johnson, and many more top-tier companies. He's the author of several books including Jump Start your Business Brain *and* Jump Start your Marketing Brain. *He also hosts a weekly public radio show called Brain Brew Idea Factory.*

Your latest book, *Jump Start your Marketing Brain* is a compendium of real-world scientific marketing results and practical ideas. And it isn't full of your own opinions—it's based on a vast amount of research.

This book was developed out of some 3,000 academic studies. Oftentimes sales and marketing are too much a right-brained emotional thing. And while you need that, I think we're working too hard. We need to figure out ways to sell more with less effort. By getting ourselves grounded more in the logical truths, we can start playing the odds instead of betting on a long shot.

I've translated the academic language into simple, practical ideas that any business professional can use.

You have a Marketing I.Q. test in this book. Here's one question: "The smartest way to consistently build sales is to: A) Build loyalty, or B) Find new customers."

Simple question.

I had a group of 200 CEO's yesterday up in Canada. And out of 200, 198 got the question wrong. They all said it was option A, "Loyalty."

But the truth is *the smartest way to grow sales is to find new customers*. Most people think it's better to protect what they've got. But the way you do that is by presenting an incredibly great product on the air. That's how you keep customers.

We've got to figure out ways to find new customers. Three separate studies confirm this: One study was looking at 6000 businesses: *The number of customers was 2.8 times more important in influencing sales volume than the amount bought per customer.*

Are you saying we in radio probably don't spend enough time trying to build Cume?

Absolutely.

> "[To grow your ratings] the number of listeners you have is 9 ½ times more important than how long they're listening"

In fact, I got the listening data from public radio, which is analogous to commercial radio. What I found was that it wasn't just 2.8 times more important. In fact, for stations which grew in both ratings and donor support, *the number of listeners they had was 9 ½ times more important than they amount they were listening!*

We've got to be reaching out. And niches of niches won't do it. Reaching out takes guts and courage because most of us just don't want to lose what we have. We've got to figure out how to get new target occasions, new target customers. That requires leadership, and it's extraordinarily difficult in today's world.

Another question from your quiz: "Most new products fail because of: A) Poor execution of sales and marketing, B) Poor product or service performance, or C) Not being a very good idea in the first place."

The answer is C) Not being a very good idea in the first place.

You can't execute your way to excellence. If it's not in the blueprints, it's unlikely that you're going to get there.

Imagine you're building a house and you say to the builder: "We don't really have blueprints. Why don't you guys just figure it out as you go along." That's about as stupid as it possibly gets, and it's what we do in business all the time.

You've really got to get your blueprints right at the beginning. Do your thinking up front—that's the job of management, the job of leadership. Don't beat up the workers for not executing, it's not going to do you any good.

Management legend W. Edwards Deming said "94% of the problems in business are due to the system, 6% are due to the workers." It's those systems—those basic ideas at the beginning—that we have to work on.

If a good idea is most important, how do you get the good idea?

There are three principles on a mental level: Two that add to ideas and one that kills you.

The first is *stimulus*, which is filling up your brain with thoughts and ideas. The problem is most of us do the same old stuff the same old way—we go to the same restaurants, read the same magazines, listen to the same stations, etc. We need to open ourselves up to new experiences, new things, and fill that mental food processor.

> "The problem is most of us do the same old stuff the same old way"

The second is *diversity*. We need to be willing to talk to people who are whacked—who are totally different from us, both "suits" and "kooks." It takes courage to bring those people together.

The third is to *reduce our fear*. Fear is a negative, and right now fear levels are at an all-time high. Like a frog dropped in water as the heat is slowly turned up, people don't even know they're close to dying—and they are.

You describe three steps to construct a marketing message properly.

Yes, first there's *"dramatic difference."*

From a listener's standpoint ask yourself the fundamental question "why should I care?" If you're promoting your show or your station, why should I care to listen to it? What makes you unique, what makes you special, what advantage do you have that no one else has?

There's a public radio station here in town that's now got billboards up—they just bought another station—and what they're saying is "We have news!" Congratulations! I'm happy for you! Everybody's got news! What makes your news unique? Is it smarter, is it deeper, etc? That's a lot of wasted money.

Tell me specifically why I should care and what makes you dramatically different.

> "Tell me specifically why I should care and what makes you dramatically different"

I believe a lot of the differences we perceive between our stations are not that dramatic. Do you agree?

Absolutely.

Most are hair-splitting differences. It's a difference to you, but not new to the world.

In the old days, we could be new to our community, but gang, as soon as Internet goes into cars we've got to be new to the world. Either that or just become a syndicated station and figure out who you're going to buy your stuff from.

People have too many choices. Today, listeners can get anything they want anywhere they want it from. It's a "flat Earth" and the Internet makes it easy to get these things. So we've got to have new-to-the-world ideas.

And a great test to know if it's really new-to-the-world is the cost of operational chaos when you change to that kind of show or format. Don't just rearrange the deck chairs. Something has to change. You can't take the same set of elements and make a miracle occur.

The future's coming whether we're ready for it or not.

In a big-time way. And I don't think people even have a grasp of the scope of what's about to happen.

There'll be a guy in Africa who becomes your competitor tomorrow because nothing's going to stop him from using the Internet to do it.

The last thing I'd want to own today is a broadcast radio station. If you're in broadcast right now you've got to think really seriously about that investment in the tower. If I owned a radio station I'd be trying to figure out what is the content I could own that I could then resell to others using my station as a test-market site—not as my primary revenue source—five or ten years from now.

> "In the future, more than ever, content is king."

In the future, more than ever, content is king.

Okay, so step 1 in constructing a marketing message is "dramatic difference." What's step 2?

"Dramatic difference" gets their attention. Now, consumers want to know what the benefit is that you are going to provide them, the *"overt benefit."*

What will they receive, enjoy, or experience in exchange for their time? What's in it for them? What will they feel?

Will they feel sexy, smarter, chill out, relax, laugh…It's primarily an experiential benefit that you're giving with radio.

Broadcasters very often advertise station features not benefits. Those aren't the same thing, are they?

No. Features are the "stuff" of radio. Benefits are what that that stuff does for you. We have to translate the stuff into benefits. In most cases, that means communicating how the listening experience is going to leave you feeling. It's an experiential benefit.

If you walk in the restaurant, have dinner, and walk out, you feel elegant, you feel romantic, you feel relaxed. How do you feel?

Likewise, you turn the radio on for some feeling. You turn the radio off—how has the listening experience changed your feeling? That's the benefit I'm talking about.

> "You need to tell listeners why they should believe you—offer a 'real reason to believe'"

You will feel more...what?

And what's step 3?

With "dramatic difference" we got their attention. With "overt benefit" we told them what's in it for them. Now, if we've done our job right, our promise is so big and unique that we need to prove it.

You need to tell listeners why they should believe you. You need to offer a *"real reason to believe."*

"We're able to offer [blank] because of...."

For example, if we claim to have the world's best news it's because we have five different media feeds we're tapping into on a regular basis so you always have the latest, state-of-the-art news.

"Real reason to believe" means convincing listeners that the benefit, as promised, will truly be delivered.

And that convincing needs to be authentic.

You can play games with testimonials and stuff, but the most impressive way of doing it is to tell the truth. Honesty and integrity wins big-time in today's world.

What's the most important thing radio broadcasters need to know right now about marketing their radio stations?

They have to figure out what their purpose is—their fundamental purpose. What is the meaningful difference that you're trying to articulate?

In the future you're going to have to internalize this: "We're going to be the best in the world at [blank]." And to fulfill that mission, that fundamental purpose, broadcasters will need to get the staff, the talent, and the production team all aligned and say we're going to be the best in the world at *something*.

For example, WUMB in Boston is a little folk radio station that carries my show. And they're number one on the web for folk music. There's a worldwide phenomenon coming out of this dinky little station. Because they do one thing and they do it great.

> "In the future you're going to have to be the best in the world at *something*"

For a bigger station you can't necessarily do one thing 24/7, but you can do one thing in the morning, one thing in the drive-time, one thing on the weekend, etc. You can assemble a package that addresses different audiences at different times because radio is not used the same way all day long. It isn't a 24/7 thing—listeners come in and out of it. So you can think about who you want your audience to be for each of those times.

HD radio is the industry's current obsession to create a future. What's your take on it?

I don't get it. I don't understand it. It seems like an interim solution, not the real solution. Why would anybody buy HD radio? It makes no sense to me.

I work with the auto manufacturers. I know those guys, and they're going to add Internet to the cars. And when they add Internet to my car, why do I even bother with HD radio?

I just don't understand it. And nobody has explained to me the purpose of HD radio. It's an interim blip.

It's the same thing with XM and Sirius. My show is on Sirius, but I'll tell you I don't get it. It's really an interim solution until we get the Internet—and then we've got everybody.

> "Why would anybody buy HD radio? It makes no sense to me."

It may sound like I'm concerned about radio. But I've built my own recording studio and I've invested in training and coaching to get better at doing radio. I think radio has phenomenal potential, but not in the form that it has today.

I think radio content has a great advantage over video. So I'm placing my money on recorded audio as opposed to video. So there's still hope, but radio is going to look different.

And what will the radio industry look like?

We're going to be delivering content all through the Internet. That's where it's going to go. Even for our little tiny public radio show we have 40,000 podcast downloads.

It'll be podcasts, one minute pieces going to cell phones, etc. Listeners will consume audio content in different formats and different places using different tools, not just a physical "radio."

It hasn't shaken out yet, obviously. But we'll see audio content transport itself to many different distribution channels.

And the person who owns the content is the one who will rule in the future.

So you have to get out of the mode of thinking about your station as a tower with exclusive access to the audience's ears?

You're a studio. Think of yourself more like a movie studio rather than a movie theater. Rather than a projector, you are the creator.

And how much cooler is that?

Differentiate or Die

Be an individual. Be unique. Stand out. Make noise.
Make someone notice.
—Jon Bon Jovi

Jack Trout is a marketing legend. Co-author of the seminal guidebooks Positioning *and* The 22 Immutable Laws of Marketing, *Trout has condensed the essence of his strategic thinking into his latest book,* Trout on Strategy. *Jack graciously took time out provide me with some of his perspective on the quality of our strategic thinking in the radio industry.*

What is the biggest strategic error you think radio stations make?

> "Most radio stations don't do a very good job of differentiating themselves from their competitors"

To me, the biggest strategic mistake in the radio business is that of not effectively selling the medium of radio. Most would rather beat up on their competitor. They should be selling radio as a primary medium as opposed to filling in the holes in the media plan.

The truth is that a marketer doesn't need pictures as the mind works by ear. While many products tend to look alike, they don't have to sound alike.

How well do most radio stations position themselves?

Most radio stations don't do a very good job of differentiating themselves from their competitors. They promote their programming as if no one else was out there with similar programming. There's a sameness, which only drives price as the main bargaining point.

And music as a differentiator is not an easy way to go because others have access to the same music.

Your book *Trout on Strategy* makes eight major points on strategy. Which one is most applicable to radio?

I would say that the chapter called *Strategy is all about Differentiation* is the most important one for radio. In that chapter, I lay out a four-step process that outlines how a radio station can build an effective differentiating program:

- Make sense in the context—your message has to start by considering what the marketplace has heard and registered from the competition.

- Find the differentiating idea—it doesn't have to be product-related, but it should provide a benefit.

- Have the credentials—don't provide claims without proof.

- Communicate your difference—Better products don't win, better perceptions win. Advertise and promote your difference.

Spread the Buzz

A kiss that speaks volumes is seldom a first edition.
—Clare Whiting

How do you capture the attention of potential listeners who are, for the most part, indifferent to your station and your messages? Mark Hughes knows. He's author of the new book Buzzmarketing, *and the guy responsible for such memorable buzz moments as the renaming of Half, Oregon to "Half.com," literally managing to put his product on the map.*

> "Are you advertising and marketing to fit in or to stand out?"

How do you define "buzz" and how can radio stations capture that "lightning in a bottle"?

Buzzmarketing means attracting the attention of consumers and the media to the point where talking about your brand becomes entertaining, fascinating, and newsworthy. First you need to seize their attention, then you need to provide people with a great story to tell or a great one to write about.

You have to give a listener the "water-cooler currency"—"hey you'll never believe what I heard on the radio this morning." You have to wrap a great story around what you want people to talk about.

What tactics should radio stations employ to generate "buzz"?

Too much of what we do in marketing is ordinary. You have to tap into the extraordinary, the unusual, the outrageous, the controversial, the hilarious, hot topics in the media, and the taboo.

You can be outrageous, as long as it's not being outrageous just for the sake of being outrageous. If you can put your own personal fingerprint on any part of a media frenzy, you can get a lot of buzz. You can build buzz by making the listener the star—for example, one of the reasons American Idol exploded is that it makes you, the viewer, the star. It's interactive.

In your book you talk about one specific example of a radio station advertiser that generated tremendous "buzz."

Totally Awesome Computers is this little company in Utah that has managed to flourish. The owner, Dell Schanze, bought some radio ads. He just walked into a studio and started recording. But it didn't sound like a commercial, it didn't sound like *anything else* on the radio. So when people heard these ads, they paid attention. Often when we hear an ad we turn off our attention button. But this was crafted as content, honest content. He would talk about lots of crazy things like French-kissing his dog. But strangely enough, those weeks where he did wacky things were his highest sales weeks. People looked forward to his content. He constantly refreshed the spots. Every single week you'd look forward to a new story from Dell Schanze. He has great products, too. *People want content, they want entertainment, they don't want another ad. And that's why he succeeds.*

> "You have to give a listener water-cooler currency"

How important is creativity in creating "buzz"?

People think that buzz is random, but it's really like baseball. You have to have a good swing and keep swinging. *Are you advertising and marketing to fit in or to stand out?* If it's the latter you're more likely to be successful.

You've got to make the big plays and not settle for most of what advertising and marketing is today—which is vanilla. But vanilla doesn't move people, and it doesn't get people talking. It's the Rocky Road that generates talk. It's the Cherry Garcia that gets tongues wagging.

In your book you talk about creating "an attractive personality" for a brand. What can a radio station learn from this?

There are a lot of conglomerates that own stations, but if you can give your station's brand a *personality* it makes a ton of difference because in today's world we really don't trust corporations, we trust people. So if you can make your brand more personal and less corporate it will increase your odds of success.

And show your warts too. If you see a little mistake here and there, you're showing that you're personal and human. Everyone makes mistakes. Instead of trying to hide everything with a polish, show those warts and people will believe you're honest.

Tell a Good Story

We want a story that starts out with an earthquake and works its way up to a climax
—*Samuel Goldwyn*

This is a story about a guy who, after founding the American Comedy Network and co-writing the pilot episode of The Nanny, *discovers his life's true calling. His name is Andy Goodman and he's now a communications consultant specializing in helping public interest groups, foundations, and progressive businesses communicate more effectively. And the method Andy specializes in is storytelling. What can radio learn from a man who knows a good tale and how to spin it?*

What does storytelling have to do with radio and the marketing of radio?

In general, people understand and relate to stories. Advertising and commercials that tell stories are more appealing than throwing listeners a bunch of numbers or benefits.

A morning show, for example, can have an ongoing story of the relationship between the host with his or her co-hosts. One of the reasons the Howard Stern show is so popular is there is a multi-character narrative going on there involving

> "Telling stories is much more appealing than throwing listeners a bunch of benefits"

his personal life and his relationship with others in the studio. People are tuning in for the latest chapter in this story.

What about your morning show? What is the relationship between these people in and around your morning show, and what is the ongoing story between them?

In News/Talk, news that tells stories is more interesting to people. Radio is a very intimate medium, it's a one-on-one medium. And we need to talk to one person at a time. One person telling a story to another is a very powerful, emotionally engaging experience. So radio is a terrific medium for storytelling.

What makes a good story?

Good stories have a point. There's a reason you tell it. By the end of the story you should know why you told it.

Somebody wants something, goes after it, runs into problems, tries to figure out a way around those problems and either does or doesn't, but ends up at a point that's different from where they started. That's the basic structure of storytelling. Somebody's got to go after something and run into a problem. Otherwise it's not a story, it's just a sequence of events.

And a good story isn't necessarily a long one. Good storytelling is concise, but colorful. One or two minutes is not too much to ask for in a single break.

How can storytelling improve a station's marketing efforts?

What's the role of the station to the lives of the listeners? If, for example, I am your listen-at-work station that means I'm your friend or companion at work. What other things can I, the station, be doing to play that role and live that story? It's not just what kind of music would be good for the background. What is the CHARACTER your station plays in the lives of the audience? What is the ROLE of your station in their lives?

How do you play that role and be part of that work story? You can't sum this up in a several-word slogan. You have to LIVE the slogan. And LIVE UP to the slogan.

Think "Creativity"

Logic will get you from A to B. Imagination will take you everywhere
—*Albert Einstein*

Tom Asacker is a top-notch marketing consultant and author of many popular branding books like Sandbox Wisdom: Revolutionize Your Brand with the Genius of Childhood *and his latest,* A Clear Eye for Branding.

What's the difference between a radio station telling the audience what we do and "Branding" ourselves?

Imagine your station is an iPod. I don't think I ever once heard Steve Jobs describe the iPod as an "mp3 player." He never said "we're the best mp3 player out there."

No one cares about your station or what you do. What they care about is how you make them feel about themselves and their decisions while in your presence.

> "No one cares about your station or what you do"

Do you make them feel special, in the know, smart, etc? Whatever it is, ask how you are making people feel while tuning you in or wearing your logo.

How do I translate that to what I do on the air every day on the air?

Understand what people are internalizing about themselves while they're listening to your station. By associating with your station I'm telling me something about me. On top of entertainment, people want to feel a certain way about themselves. How can we help them do that?

When we walk down the street with a coffee cup with a green logo on it that says "Starbucks" we're giving out a message about who we are. When I listen to your station or wear your logo I'm doing the same.

You say consumers—listeners—gravitate towards things that boost their self-esteem, their sense of self. How do radio stations tie into that?

First, gain an understanding of what that sense of self is and appeal to it with content. Or have an attitude, an opinion, a worldview, and express it creatively so people are drawn to you because they share the same sense of the world, that same sense of self.

You say we should forget about selling our unique position and instead uniquely express that position. But what about the logical appeals that we tend to rely on, the 'Lite Rock, Less Talk' or 'Best Mix' appeals?

> "Don't think 'Positioning.' Think 'Attitude,' think 'Creativity'"

People don't make decisions logically. People are not rational creatures. The iPod didn't sell a unique position. BUT it uniquely expressed its identity in the design of the unit and the ease of use of iTunes. You don't focus on the "it." Instead, discover the attitude and the point of view, and draw the audience in, engage them. With Carl's Jr. it wasn't the hamburger, it was Paris Hilton.

Companies like Apple and Starbucks don't have positioning lines because they're not interested in comparing themselves to their competition, they don't choose to run in the same race that competitors do. They decided to break out of the pack. Starbucks is NOT a kind of coffee shop—it stands apart from the rest. Companies like this don't think positioning, they think attitude, they think creativity.

Your book talks about the importance of developing a strong story. Can you think of a radio station that has a great "story"?

I can talk about my own tastes. NPR, for example, has a story that tells me about me (which is what all good stories do). They're telling me that I'm well informed, in the know, smart, in tune to what's happening. That station helps me fulfill that particular way of being while I listen to it.

Of course, that's not how I am all the time. I'll listen to WEEI in Boston for a different story, one about being sports-savvy and a Boston native and a die-hard sports fan.

A station has to be true to its particular worldview. Whatever that attitude is. Because we want to know what attitude to expect when we turn on the radio.

How to be Funny

I never forget a face, but in your case I'll be glad to make an exception
—Groucho Marx

What does it take to make a truly entertaining morning show? Anne Libera knows. She's a director at Chicago's illustrious Second City, ground zero for much of the world's finest comedy talent. Her new book, The Second City Almanac of Improvisation, *is full of pointers and exercises that can make your morning show sparkle.*

Take some lessons from the training ground of talent like Joan Rivers, Robert Klein, John Belushi, Dan Aykroyd, Bill Murray, John Candy, Martin Short, Gilda Radner, Tina Fey, Bonnie Hunt, Mike Myers, and many, many more.

> "Comedy is an equation: Truth plus pain plus a certain level of distance"

Why would Improv training make a better radio morning show?

Improvisation and entertainment-oriented morning radio are the same thing. We are all just making stuff up in the moment. And one of the biggest blocks to success in both media is the fear of failure that manifests itself in judging each word before we say it—this keeps us from allowing our creativity free rein.

Many of the improvisational basics are geared towards specifically building self confidence, self awareness, and unleashing creativity. A lot of that boils down to giving yourself something specific to focus on—what we call a *point of concentration*. For most of us when we're performing, that point of concentration is something like "be funny" or "be brilliant" which just invites awkwardness and paralysis. Improv training gives us the chance to play with a variety of points of concentration and take risks in the safe atmosphere of a workshop setting—all things that we can then take and use in the more stressful situation of being on air.

Also, radio is a personality driven medium. Improvisation helps to hone point of view and personal voice of the sort that creates a strong on-air persona. You create a stage (or radio) persona by discovering the truth of yourself and heightening it.

Most radio morning shows are pushed to be funny. Why do you feel it's more important to be "true"?

I can't say enough that I think that the absolute best comedy—the stuff that makes us all laugh from our gut happens when the truth is spoken—especially if it is a truth that everyone has been thinking but were afraid to say.

It's what makes Howard Stern's work brilliant—even at his grossest. And at the opposite end of the spectrum Garrison Keillor (whom I grew up listening to as the morning DJ on Minnesota Public radio) points out softer but no less universal truths in Saturday monologues.

What "rules" of Improv do you think best apply to morning radio?

The principle of "Yes, and…" is key for radio—accepting that whatever is said, whatever happens is a gift—it's supposed to happen and then building on it. "Yes, and…" allows you to keep going no matter what happens because you never stop accepting what has just happened and taking it to the next level.

> "Real comedy comes out of truth"

Here's another principle: *Real comedy comes out of truth.* Character is a filter— it's what a certain person sees, hears, pays attention to or blocks out. And how that information is processed is an enormous part of what makes us laugh—we recognize both what was inputted and what came out as recognizable human behaviour. And that's what really makes us laugh.

I tell my students that comedy is an equation—truth plus pain plus a certain level of distance (it's not funny if it happens to me). There are some morning shows where the banter never hits any actual level of truth or pain—its just a little truth and a whole lot of distance.

On the other hand, I also tell my students that the pain in that equation can be what I call the "ewww" factor—gross out or shock value or cruelty. You've gotta have a little of it—and I'm the first to admit the perverse pleasure in the discomfort of others (*The Office* and *Curb Your Enthusiasm* come to mind). But I'm afraid that there is a tendency in morning entertainment radio to rely on that exclusively as opposed to honestly letting a touch of reality intrude and finding the humor in the pain of our shared human condition.

I guess what I mean is that I'll laugh at a extreme gross out joke because it makes us uncomfortable but I don't want to spend the entire morning there.

Market Across Generations

Life would be infinitely happier if we could only be born at the age of eighty and gradually approach eighteen
—Mark Twain

"What youth deemed crystal, age found out was dew," wrote Robert Browning. But just try telling that to your young and youth-fixated agency buyers if you're selling a station which targets listeners over 35, 45, or—God forbid—50.

David Wolfe is on a mission to change that unfortunate bias that has so many of our older-targeted stations in its grip. Wolfe is a speaker, marketing consultant, and co-author of the book Ageless Marketing. He also writes a terrific and influential online blog, http://www.agelessmarketing.com.

Lots of radio stations have audiences that skew 35–54 or 35–64. They often find that buyers are biased against their stations for this reason, regardless of how healthy their ratings are. What advice would you give these broadcasters?

> "The largest population and greatest pending power clearly lies with the 35-plus set"

Those stations might consider coalescing into an association with a suitably descriptive name, and develop a marketing campaign to persuade advertisers that the big payoff in markets today lie with those age groups.

The numbers support that, but advertisers—aided and abetted by the young people in marketing—haven't fully focused on the fact that the largest population and greatest pending power clearly lies with the 35-plus set.

How big is this oldish market? How much money is there? Do they spend it? And if so, why don't advertisers care to reach this audience with the same zeal they reserve for 18–49s?

Consider the statistics:

- A Baby Boomer turns 50 every 7.5 seconds
- In 2005, half of all people between the ages of 50 and 74 are Boomers
- The size of the 50+ population will more than double over the next 35 years
- The 50-plus set makes up 35% of the population
- The 50-plus set controls 77% of privately held financial assets and has 57% of the discretionary income

Though spending peaks around age 50, the combination of stalled population growth in the under-50 population and the explosive growth in the 50-and-older population means no room for sales growth in the adult under-50 age group but *huge* sales growth potential in the 50-and-older age group.

> "The secret to intergenerational marketing is to project universal values"

As to why advertisers have been ignoring these huge and growing markets, I think it's largely a matter of negative stereotypes of older people as consumers and a sort of Newtonian-like principle by which a belief in place, like an object in place, tends to stay in place. Most people simply don't like to change even when by any objective reckoning they should.

Is it possible to appeal across generations if your natural market is older? How can values-based appeals bring in younger listeners without sacrificing the base?

Great questions. Intergenerational marketing is more important than ever before because the age groups that traditionally spend most in the consumer economy have stopped growing. By shifting to an intergenerational marketing approach you extend your brand's reach.

New Balance's spectacular growth over the past 15 years—when it moved from #12 sneaker maker to the #3 position—was fueled by its shift from age-segmented marketing to intergenerational marketing—what I call *ageless marketing*.

Certain values are age specific, such as the social values that confer on adolescents their herd-like behavior. However, other values are universal, such as self-reliance, love, fidelity in relationships, and of course, God, flag, motherhood and apple pie. *The secret to intergenerational marketing is to project universal values.* Hallmark has been doing this for years. Young and old alike will reach for their Kleenex when watching a Hallmark commercial. That's what happened with one of Coke's all-time most successful television commercials—Mean Joe Green—that ran in 1980. People remember it with astonishing accuracy 25 years later!

Energize your Brand Evangelists

I'm the person most likely to sleep with my female fans, I genuinely love other women. And I think they know that
—*Angelina Jolie*

Word-of-mouth is the most powerful form of marketing under the sun. How do you get listener tongues wagging and, in the process, move the needle for your station's ratings? Dave Balter knows. He's the creator of Bzzagent.com and a founding member of the Word-of-Mouth Marketing Association. He is also co-author of the upcoming book Grapevine: The New Art Of Word-Of-Mouth Marketing.

> "85% of agents have never redeemed a single point!"

What is Bzzagent.com and how can you help a radio station?

Bzzagent.com is a word-of-mouth marketing and research firm. We have over 100,000 volunteer brand evangelists or "agents" who take part in word-of-mouth programs for products and services and brands and share their opinions with others as effectively as possible.

There are lots of people who have opinions about the stations they listen to. Our goal is to identify these people, organize them, help them communicate more effectively with others, and help the station to understand how listeners are discussing it.

The goal is to understand what they're saying and to get them more involved in communicating better.

What are the benefits of Bzzagent.com to your clients?

Agents get to try a product or service, then they go out and create word-of-mouth. They come back to us to fill in a "buzz form" which details their communication and what happened. We then write an individualized reply to each one—no automation. We get about 7,000 of these submissions each week.

For the customer, there are two outcomes:

First, to engage consumers and build a community of people talking to others effectively. For example, that would help a radio station acquire more listeners.

Second, we're gathering data about how people are actually communicating. In each report we can see how many people are communicating, when they're communicating, what they're saying, what age groups, etc. This not only allows clients to monitor the conversation, it also helps them see opportunities to change their marketing, their product, or their service to make it more effective.

> "It's not all about incentives. Recognition is a pretty important factor"

The way people communicate is in motion. People come to us to see how consumers are changing their communication and how they're influencing others. That knowledge, awareness, and access to understanding is the main value driver.

And what's the benefit to the members, the "Bzzagents"?

We intended for members to generate word-of-mouth for points they can redeem for brand-associated rewards. But *as of last month 85% of agents have never redeemed a single point!*

Naturally, we looked into this. And agents told us they participate for many reasons besides points. They want to be the first to know, they like to talk about new things, they're fascinated that they have a dialogue with a brand, they're grateful that people are listening, they appreciate the power of their opinions.

Lots of radio is based on incentives, such as contests. What does that discovery imply about the way radio lures new listeners?

It's not all about incentives. Recognition is a pretty important factor. I think a lot of people overlook the value of being embraced by the brand. Sure, perks are good—but it's more important for listeners to feel they're part of the system.

So for radio, how do you get listeners to feel they're part of the solution and not simply a "targeted customer"?

If you ran a radio station "Frequent Listener Club," how would you use it?

> "The light loyals—not the heavy listeners—are the people who generate the most effective word-of-mouth"

First, figure out ways to let listeners communicate to you what they're doing, how they're sharing their opinions, then be responsive to each one. It's a daunting task, but it's what will drive people to become more involved.

Let people feel on the inside and feel they're really communicating with the brand and are part of the solution—not just a "target." How you treat these people will really make a difference.

Are all "Frequent Listeners" created equal?

We expected that heavy listeners—the influentials, the mavens—were the most valuable people from a word-of-mouth perspective.

But we discovered that these influentials didn't talk any more than most other people. The heavy loyals have already influenced their network of friends—and they felt they "owned" the brand already and thus didn't talk about it much. The experts, meanwhile, had very short word-of-mouth windows—they'd only talk about something as long as it was new, then they'd move on to the next new thing.

The "light loyals"—the people who live on the bottom of your database list—these are the people who generate the most effective word of mouth for the brand. They have a network around them that hasn't been influenced and they're not basing their identity on how cool they are or how much they know about your product.

This is the group of people driving the most value. Look at the group that you often overlook—not the biggest fans but these "light loyals"—these will be the most passionate evangelists that will drive the most long-term results. Don't limit yourself to targeting the heavy listeners.

Why are "light loyals" more powerful word-of-mouth generators? Their motivations are different. If they're in your club it's to be involved with the station. They're not in it for the perks. And these are the listeners with "fresh ears" around them.

What are the best ways a radio station can generate word-of-mouth?

Don't think of word-of-mouth as some special kind of marketing medium. Word-of-mouth is something that's happening all around you every day. In one study, 14% of every conversation had something to do with a product or service.

There are opportunities for your listeners to talk about you all the time. Know the "triggers" and use them. Know the targets—the types of people you want to talk to. You need to make people conscious of the right way to communicate, because it's happening all around you anyway.

The Radio Marketing Playbook

The other teams could make trouble for us if they win
—*Yogi Berra*

John Zagula is the co-author of the business bestseller The Marketing Playbook *and co-host of its affiliated (and highly regarded) marketing blog, http://www.marketingplaybook.com.*

This is one of the best and most practical marketing manuals I've found. That's why I asked Zagula, a former Microsoft marketing executive, to apply some of his strategies to the radio industry in this Q&A.

> "Positioning is the argument that you are different, relevant, and preferable"

Radio stations often find themselves in direct competition—this is equivalent to what you call the "drag race" play. What are the rules of the drag race play and how can a radio station use these rules to win the race?

A drag race is a strategy of direct confrontation. Even if you find yourself in direct competition you should always make a conscious decision whether you want to confront your competitor directly or not.

Do not enter in a drag race unless you both have a lot to gain by doing so and you have what it takes to win. What do you need to do to win? Go for it—all the way. Steer the race course in your favor.

Be direct and relentless in your comparisons. Demonstrate momentum—in whatever terms you can. Once it looks like you're actually gaining it gets much harder for the other driver to reverse this perception (this implies that it often makes sense to drag race if you are number two—comparisons and PR will more likely shine on you and improve your status relative to the leader who is put on the defensive).

What is the difference between "positioning" and "messaging"? Many folks in radio seem to think they're the same thing.

Positioning is simple. It is the argument you have to make that you are different, relevant and preferable.

Messaging is the translation of that argument into words that actually grab the attention of your target.

Positioning is long term. It is the basic argument you have decided to make. Messaging can change. It can be tested to see if it resonates and actually gets this argument across.

You have an equation: P>B>F. What does that mean?

This is simple. When you go out and turn your lovely positioning into messaging it's best to support your claims and stories. And the best support works like this:

PROOF is stronger than BENEFITS are stronger than FEATURES.

What do I mean? Well it may be good that you are on twenty four hours or reach 27 cities or have dog and cat health tips (FEATURES) but why does that matter to your target? What do they gain from you having those things? More audience to advertise to, news whenever they need it (BENEFITS)? And why should they believe you? How about because thousands/millions already listen, you're the fastest growing, leading experts agree, etc. (PROOF).

Become a Cult Brand

I wonder if other dogs think poodles are members of a weird religious cult
—Rita Rudner

B.J. Bueno is co-author of the book The Power of Cult Branding. We toss around the term "brand" in our industry to apply to every station, but as B.J. points out, there's a big difference between a "blah" brand and a truly great "cult" one. Here's B.J.'s twist on cult branding for radio.

Your book highlights seven rules of Cult Branding. Which are most important for radio?

> "If Radio wants to win heart-share with their customers they will have to take risks"

First, consumers want to be part of a group that's different. Or best said for radio, listeners want to hear a station that is different.

It is unfortunate how many quick followers live in radio, and today formats are copied so quickly that the original ideas are not allowed to take form and find their audience. The world of radio is full of sameness and in today's competitive environment radio can't afford to build a station where the listener feels that his or her station has no substitute, no equal, no match. We humans love to belong, but belonging is not enough—we want to be in with the group that is different than the rest. This is a great opportunity for radio.

Second, cult brand inventors show daring and determination. Yes they do. And here is where we are back to square one in radio. If radio wants to win heart share with their customers (both advertisers and listeners) they will have to take risks. All great cult brands have taken enormous risks without looking back at 'what if's.' Radio has become risk averse, but today the biggest risk is not taking a risk at all. Radio must find inventors that are daring and determined.

What do you think radio stations really need to do to achieve the status of cult brands?

Radio must follow its bliss. When was the last time something new and innovative was done in radio? Where are the people who are challenging the status quo of radio?

All cult brands acted as families for its members. Giving them a place they can call home. All cult brands loved their customers and did everything to empower their group. World Wrestling Entertainment creates real time dramas that are played out and result in the victory of the one the crowd wants more. Sometimes they create long stories and struggles that play themselves out over time. Where is the drama in radio? Where are the loyal fans that are involved in the family? I, to my disappointment, have even heard some people complain about P1's because the are always calling and want things. Wow! I am not sure that is a problem. When the listeners wins everyone wins.

What do you think radio's biggest mistakes are as we try to build Cult Brands?

That's easy: Please, radio, take a risk. Start to reward failure and punish people who are cash dispensers. In times of change you can't afford people who don't want to take ownership of problems and think creatively.

Sell the Dream

Nothing great was ever achieved without enthusiasm
—*Ralph Waldo Emerson*

Guy Kawasaki knows how to fix what ails your station. Formerly chief evangelist at Apple Computer and managing director at Garage Technology Ventures, he's the author of eight books, including some of my favorites: Selling the Dream, The Macintosh Way, Rules for Revolutionaries, *and his most recent:* The Art of the Start, The Time-Tested, Battle-Hardened Guide for Anyone Starting Anything.

> "The starting point of evangelism is a great product"

If you ran a radio station, what would you do to make it stand out?

I would hire knowledgeable, vibrant, and opinionated personalities. Not shock jocks that want to create a ruckus by being controversial, not someone who stopped playing baseball in the 6th grade who's now telling you why Barry Bonds is over-rated, but people really in the know. Greg Kihn on KFOX in San Jose epitomizes this. Because he was a rock musician, he really knows what he's talking about. He's not some wannabe or never-was that's trying to prove himself. He gets great guests—how many people interview Paul McCartney on their show?

On a conceptual level, draw a graph with the vertical axis measuring the station's ability to provide unique content and programming. On the horizontal axis, measure the desirability of the programming to people. So if you're high on the vertical axis, only your station can provide something. If you're far out on the horizontal axis, people really care about what you provide. The goal is to be high and to the right—like George W. Most radio stations are low on the vertical axis and moderately far out on the horizontal axis. For example, yet another station providing rock music. Greg Kihn puts KFOX high on the vertical axis. How many other rock radio stations have a bonafide rock star as an announcer?

What are radio's biggest marketing mistakes, from your perspective?

I'm on the outside looking in, so for all I know, they work, but the giveaways and promotions seem over-used to me. If you're like KKSF in San Francisco and you can give away a trip a day to Hawaii, then there's some real branding and loyalty power to a promotion. But the sporadic sweepstakes and prizes seem superficial. I'm a romantic. I believe in great "product" where product for a radio station is its music, commentary, or announcers. Most promotions just don't excite me as a marketer.

Also, a radio station can only stand for one thing in the public's mind. To shift gears, no pun intended, let's look at cars. Volvo equals safety. No matter how sexy Volvo tries to make its cars, it will still stand for safely. KKSF is Smooth Jazz. KFOX is Classic Rock. Stations need to realize they can only stand for one thing and then plant a deep stake in the ground to own that category like Volvo owns "safety."

> "A radio station can stand for only one thing in the public's mind"

How do stations energize their listeners and transform them into "Evangelists"?

The key to evangelism is a rule that I call "Guy's Golden Touch." It goes like this: "Whatever is gold, Guy touches." The starting point of evangelism is a great "product." After that, it's easy. You go out and purposely try to build a community. You unabashedly ask for help. Many organizations think that they shouldn't ask—either because it's a sign of weakness or an imposition. If you have a great station, it's neither.

Then you need to give them the tools to evangelize your organization such as an explanation of your programming, background of announcers, calendar of events, bumper stickers, whatever. You also make them feel special with get togethers, concerts, t-shirts, stickers, CDs, and MP3s. You make your staff press flesh with them. You hire someone to empower them and make them happy. Finally, you respond to their comments. They are your best salespeople. You should cater to their desires and listen to their feedback.

Stand for Something

You have enemies? Good. That means you've stood up for something, sometime in your life
—*Winston Churchill*

John Moore spent eight years in corporate marketing at Starbucks. Now he's director of national marketing for Whole Foods Market, the world's largest natural and organic supermarket and the co-author of the marketing weblog BrandAutopsy (http://www.brandautopsy.com).

What can radio learn about marketing from Starbucks and Whole Foods?

1. Make the Common Uncommon

Whole Foods made the common grocery shopping experience uncommon by focusing supremely on natural and organic groceries that taste good and make one feel good. Starbucks made the common cup of coffee uncommon by focusing on higher-quality beans and a higher-quality experience.

> "Stand for something, not everything"

I view radio as being a commoditized experience. From station-to-station, parity among radio broadcasters is prevalent. They all air the same promotions, play the same commercials, and run roughly the same programming style. For the most part, very little differentiation exists between radio stations despite the myriad formats. Radio has to make the common radio listening experience uncommon by not following preordained rules that say the only way you can give away concert tickets is to the 10th caller. Or, that all station IDs must be over-produced, sugar-sweet, and harmony-heavy. Being uncommonly good at the common means breaking a few rules in the process—doing so will help radio stations become meaningfully unique.

2. *Stand for Something, Not Everything*

Whole Foods stands for a natural and organic approach to food, they do not stand for overly-processed, mass-produced goods that conventional grocers stand for. In fact, Whole Foods has a well-defined "Quality Standards" checklist that all products must pass before going on the shelf. These dogmatic "Quality Standards" not only define what makes Whole Foods remarkable, it also safeguards them from encroaching competition because other grocers cannot stomach the idea of sacrificing the many for the few.

To differentiate itself from like-minded competitors, radio stations could develop and dogmatically follow their own meaningful "Quality Standards."

3. *Be Mission-Bound*

Transcending commoditization requires more than having a mission statement, you have to live a mission. Whole Foods and Starbucks are both mission-bound companies. For Whole Foods, the mission is about changing the way the world eats. And for Starbucks, the mission is about getting folks to enjoy a more rewarding and inspiring coffee experience.

> "Stale programming ideas will only make it more difficult to become uncommonly good"

Does your radio station go beyond having a mission statement to living a mission? Does your radio station seek to positively change the lives of its listeners?

4. Fight the Tired and Trite

The tried and true have become tired and trite. For example, why do stations endlessly repeat monikers like KISS 106, B 93, and Magic 95? I know these are mnemonic devices for listeners with Arbitron diaries. But if you are going to repeat your station ID hundreds of times a day, shouldn't it personify the essence of what makes your station remarkable and not be used merely as a gimmicky mnemonic tool? After all, what does KISS, B, or Magic mean anyway? Second, why do stations recycle tired and trite phrases like "All Hit. All New."? This line has been used so often that it carries little or no meaning. The same goes for trite features like "Smash it or Trash it" or "Top Five at Five." Relying on such stale programming ideas will only make it more difficult to become uncommonly good.

All of the above are "best practices" that have been shared from station to station for decades. Sharing "best practices" is great, but when a best practice becomes tired and trite it becomes a "worst practice." Radio stations need to reduce "worst practices."

5. Have a Reason for Being

Because the radio dial is crowded, it is imperative that a station has a clearly defined and articulated reason for being. To define this, a station needs to answer two questions:

1. What makes this station special (i.e. remarkable)?

2. Why should a listener care?

> "Have a reason for being"

Once a station has defined its reason for being it has defined its brand. With its brand defined, the station needs to articulate their reason for being in everything they do. All employees must become the marketing department. Every decision must be based on an understanding and appreciation of the station's brand. And why should listeners care? The answer must be genuine and be truly relevant to the listener. Listeners today are far too savvy and can quickly see through bogus and patronizing answers.

6. Challenge Your Promotions Department

To succeed as a promotions-driven radio station, I believe the advertiser's objective must become the station's objective. For example, if a car dealership's goal is to sell 40 new cars during a three-week promotional flight, then the station's objective is to do everything within its promotional power to help the dealership meet or exceed that goal. Ultimately, I believe advertisers buy results more than ratings.

Here's an idea: When giving away concert tickets, don't ask listeners to be the 9th caller to win. Instead, I would drive traffic to a local advertiser's business and have the lucky listener(s) pick up their free tickets there—at the retailer. For call-in remotes, I'd insist on crystal clear sound (not the typical static-heavy sound other stations find acceptable) and my station would always send a recognizable on-air personality (never an intern).

For local and regional advertisers who lack the financial resources to create great radio creative, we would design dynamic on-air station-driven promotional campaigns that will generate results far better than a cheaply produced radio spot.

Don't Be Average

I'd rather have a moment of wonderful than a lifetime of nothing special
—*Anonymous*

He has written Permission Marketing, Unleashing the Ideavirus, *and* Purple Cow, *some of the best-selling marketing books of all time, and now Seth Godin turns his sights to radio. This interview followed one of Seth's more recent books,* Free Prize Inside.

What do you mean by "Free Prize Inside"?

> "Radio is afraid to take things to the edge and make themselves too remarkable"

A "Purple Cow" is a product or service that's remarkable. And a "Free Prize" is that thing about your product or service that's worth remarking on, seeking out, and buying or listening to. It's not about what we need, it's about what we want. A "Free Prize" rarely delivers more of what we're buying in the first place. It delivers something extra, something remarkable.

Casey Kasem, believe it or not, was a great example of how this can work successfully. The top 40 are virtually the same everywhere, but it's the writing and the way he continually met expectations that made people pick him. Same with the *Car Talk* guys. In my town, the local non-profit has a show that competes with *Car Talk*, but they foolishly believe the show is about fixing your car.

What do you think is radio's biggest marketing lost opportunity?

Too often, radio is afraid to take things to the edge and make themselves too "purple"—too remarkable. They fear turning off the mass audience.

Well, time to get used to being marginal. Time to hone your excuses and polish your resume. In an XM, webcasting world, why on earth would I pick the average station when I don't have to? There are a thousand things you can do that'll get you to an edge. Just don't be average.

How can radio find that "Free Prize Inside"?

Radio believes that tricking the Arbitrons (repeat those call letters. Over and over and over!) and appealing to the average masses ("In the next hour, we'll play Norah Jones' new hit six times!") is the way to grow the numbers. Sure, in the short run, it is. But it doesn't get you real growth, growth that could change the rules and get you in a position of dominance.

Quick question: how many email addresses of listeners do you have—with permission to use them? If you mailed them all, could you get 10% to do what you ask them to? If not, why not?

If I were in radio I'd adopt a posture of always looking for the new style, always seeking the edge and the Free Prize.

Don't bet it all *once* and have to get it right. Bet *all the time* and sooner or later you will get it right.

Impact is Cheap

I don't want for visibility, but people seem to forget pretty easily
—Gary Coleman

It was one of the most important things learned in Mercury and Point-to-Point Marketing's recent research project on how to make your marketing more effective: The importance of a strong Public Relations plan in your war on listener indifference. "From friends" and "from a newspaper or TV story" were two of the ways listeners most wanted to find out about a new station. And strong PR can get you both.

Richard Laermer is head of RLM Public Relations and author of the business bestseller Full Frontal PR: Getting People Talking About You and Your Business or Your Product. *Richard was also one of the stars of the TLC network series* Taking Care of Business.

What is public relations and what can it do for radio?

> "Impact is cheap. Failure is expensive"

PR is not banner-hanging, it's not parking a van at a remote, it's not sending out news releases, and it's not being the 100th of 101 booths at a community festival. PR is about GETTING BUZZ.

Buzz—word-of-mouth—is what drives listeners to new stations. Buzz comes in part from that segment on the TV news or the print pieces that tell your station's story. Buzz comes from friends or through media, not through advertising. Why?

Because buzz is not sponsored, buzz comes from an independent third party, buzz is authentic and credible. Ask anybody who reads papers or watches TV: Do they believe the news or the ads?

As an industry, radio doesn't do much public relations, does it?

Public relations and radio are oxymoronic. I have a huge love for radio, but radio does nothing to promote how great the content is, how easy and free the technology is, how much it's steeped in tradition, or how great the personalities are—some of them, anyway.

But here's the problem: Somebody in their 20's who's listening to Internet radio is not going to bother with call letters—they think it's like their father's Oldsmobile. It's not because radio's not good. But radio doesn't seem to change with the times. Having Dr. Laura, Rush Limbaugh, and Howard Stern stand for an industry was interesting ten or fifteen years ago. But today, if you're 21 years old, why would you care about these old farts?

What can our industry do to be relevant to new generations?

> "The essence of PR is being new"

People in radio have to learn from the TV people and stop their navel-gazing. Everybody I know in radio is always talking about the same-old same-old. Why not get out there and talk to the media, talk to the influencers, start a revolution. Without being cliché, *radio should start a movement* and make people knowledgeable about radio in a way they're not right now.

If everyone in radio is doing all the same thing all the time over and over, then you're teetering on a precipice. Anything could happen. Satellite and Internet could take over because they've got the ideas. Even Howard Stern isn't dangerous anymore because he's an old idea.

Is having news the essence of PR?

The essence of PR is being new. We call it cultural integration. No matter who you are and what you're doing you and your product can be part of the news of today.

I've always thought that two things were missing from radio:

First, personality—really fun, interesting, sometimes shocking, a little offbeat, colorful, and surprising.

Second, talk about something that nobody else talks about. What you hear now is all the same. It's boring, it's mindless. You can't be mindless to people who are multi-tasking, like those 20-somethings I was talking about before.

How can radio stations do a better job at public relations?

Distill your message to its essence. Make it unique and differentiated—that makes it easier to promote to the world. If you want to get media you've got to make yourself media-worthy. And piggy-backing your logo on community or sporting events and concerts is doing nothing to make you media-worthy. In radio *visibility* is too often confused with *impact*.

There are many things a good PR firm can bring to your station. They can help you clarify your message, craft an angle that's appropriate for each media contact, introduce you to the reporters you need to meet (note: It's NOT just the reporter who covers the radio beat), and follow through with those reporters. It's about relationships, not mountains of meaningless press releases.

Think about that next time you're deciding how to spend your marketing dollars. After all, impact is cheap. Failure is expensive.

What's in a Name?

I grew up in Hollywood. Saying my name here is like mentioning Ford in Detroit
　　　　　—Tony Goldwyn

Virtually no topic bogs down a strategic meeting more than this one: What do we name our new station? Igor International is one of the world's foremost naming and branding agencies. And Igor's creative director, Jay Jurisich, has much to say about radio and the name game.

Your firm specializes in naming products, services, and companies. How would you assess the names you see applied to radio stations?

> "A bad name is a name with no soul"

Most radio station names either play-off the call letters or have some experiential aspiration, such as "POWER 106," "LIVE 105" or "HOT97." Some achieve a unity of call letters and evocative name—KFOG in San Francisco, for instance, invoking the city's most noteworthy climatological feature as well as a stoned rock concert state of mind.

Many radio station names are basically mnemonic devices for remembering the call letters—stations like KROQ in Los Angeles ("K-Rock") or New York's WHTZ ("W-Hits"), and some even manage to turn the mnemonic into a brand, as did San Francisco's KLLC, known as "Alice," a name that goes beyond the call letters to effectively evoke its "chick rock" brand identity as well as referencing Lewis Carroll's famous Alice (their in-studio webcam is called the "Looking Glass") and the lyrics of "White Rabbit" by Jefferson Airplane ("Go ask Alice...").

A growing trend, I think, is that more and more radio stations are beginning to realize that there's no law requiring them to be named after their call letters, so you get stations like San Francisco's KSAN calling themselves "The Bone," a name related more to their hard classic rock format and brand identity than their call letters (which, typically, just relate to the local area). When a station has an evocative name, it has more than just call letters or a handy way to remember the call letters—it has a brand. And since radio is now such a competitive big media business, brands are more important than ever. So The Bone's listeners are called "Boneheads" and KFOG's are called "Fogheads," and all kinds of promotion is done playing-off the names.

However, even a name like "The Bone" is being copied: once by a station in Texas with the exact same format, logo and positioning, which, because it's owned by the same parent company (Susquehanna Radio Corp), is understandable; less so are the two other classic rock format stations calling themselves "The Bone," one in North Carolina and one in New York, with different but very similar logos. Perhaps they've all worked-out some sort of an agreement regarding trademark, but purely from a branding perspective so many stations with the same name and very similar positioning demonstrates that the brand is not very unique. But since broadcast radio is still by and large a regional business, such cloning is not as egregious a brand faux pas as it would be in other industries. Not yet, anyway.

> "You want a name that will make people ask 'What's going on here?'"

What are the most important things a radio station should be mindful of in naming itself? How do you distinguish between "good" and "bad" names?

A powerful brand draws consumers to it like moths to flame. Apple could enter the radio market with an "Apple radio" product and immediately generate a lot of buzz; in effect they've already done this with iTunes for the MP3 market. Virgin is already in the radio business, but over the Internet rather than on FM. Their UK channel is called "Virgin Radio" and their US channel is called "RadioFreeVirgin" As with most things Virgin, they have done an excellent job of branding, and break many of the usual taboos, including that of the single, brand-reinforcing tagline—RadioFreeVirgin serves-up a new tagline every time you load a page, including such unconventional tags as "Sometimes I can't feel my feet;" "Never tested on animals;" and "Listen at your own risk." Virgin consistently does the unexpected in every market they enter, backs up their marketing efforts with real actions that benefit consumers, and in the process generates fierce customer loyalty.

Of course, the fact that both Apple and Virgin have well-established brands with a high cool factor gives them a big advantage when they enter new markets; but remember, they were once small companies. The important thing is to think beyond the boundaries of the current market sector to create a brand that consumers can form emotional attachments to, and to begin life with a name that can be just as powerful ten or twenty years down the road, no matter what path your business follows. That's what all great names do, no matter the industry.

A bad name in radio, as in any industry, is a name with no soul, that the consumer can't get emotionally attached to, and therefore will not remember. You want a name that will make people pause, even for just for a few seconds, scratch their heads and think, "what's going on here?" When that happens, they're

> "I'd avoid all those 'Star,' 'Kiss,' and 'Hits' words—they're just deadly"

engaged, and once you've achieved engagement the listener is ready to give you a chance. That's all you really need the name to do for you—beyond that, you have to back up your name and messaging with meaningful actions, you have to demonstrate why you are different and why consumers should care about you at every opportunity; this is what is meant by "branding." So while your name can get you noticed, and even remembered, it's your brand is what will make people want to remember you and come back to you; the prerequisite, however, will always be a name that is capable of becoming a strong brand.

Most radio stations copy their names from like-formatted stations in other markets. There are scores of "Star" and "Kiss" and "Lite" and "Mix." Is this good or bad?

As I've said, you need to differentiate yourself from your competitors, otherwise you'll be lost in the crowd. So in the radio space, I'd avoid all those "Star," "Kiss" and "Hits" words, they're just deadly. Then again, for those stations that are selling a homogenized, least-common-dominator type of "light urban contemporary easy listening" music, perhaps a watered-down, generic name is appropriate. Maybe that appeals to their audience, just as "The Bone" appeals to the classic rock crowd of any given city.

Why is a radio station's name so important, anyway? Can a great product overcome a mediocre name?

In theory yes, but the full answer is that a great, successful product always could have been even more successful if it had a great name rather than a mediocre name. Again, it goes back to the branding, which is intimately related to a company or product's messaging at every point of contact. If the business proposition is that a given new company or product (or radio station) is "revolutionary," "totally unique," or "unprecedented" (which these days nearly everything claims to be), but it has a boring name that is just like all its competitors, then it is undermining its core values. Consumers will pick up on such brand disjunctions, consciously or unconsciously, and it will become a barrier between them and complete emotional engagement with the brand.

A great example of this playing-out right now is the success of Southwest Airlines vs. the success of JetBlue. Before JetBlue, Southwest was very successful with a generic name simply by beating the other airlines on price. Then JetBlue came along and matched Southwest's low fares. But that wasn't all it did. JetBlue made many smart branding decisions, beginning with its name, including adding amenities such as leather seats, each with its own TV screen, great service, and the counterintuitive masterstroke of actually taking away airline meals, which nobody likes anyway. Now JetBlue has a fanatically loyal following, and Southwest, which never created true brand loyalty among its customers, is losing them in droves to JetBlue, while discount newcomers like Song and Ted are forced to imitate and play catch-up.

Radio, however, is a bit of a special case because there is a small, finite number of stations competing in a given market, so a great brand name is of less importance, at least for the moment, than in most other industries. But "for the moment" is the important qualifier here, because things are changing so quickly with technology such as webcasting, MP3s and satellite radio, that in five or ten years we may have an entirely different definition of "radio," and a strong brand may be crucial. So I think it's best to err on the side of caution, and create a memorable brand even if you think, today, that you don't need one to compete. Chances are great that soon, you will. Just ask Southwest Airlines.

Naming aside, if you owned a radio station in the Bay area, how would you approach its branding (I know this depends on the format, but I'm curious about your approach in general)?

> "Strive to capture the listeners' imagination"

If it were my station, regardless of market or format, I would burn the template that all other commercial stations follow, since I don't understand the point of being a "me-too" company in any industry. Beginning with the name, which would be different, evocative an memorable, down to every last detail of the brand: the music (much more eclectic), the playlists (extensive), the djs (intelligent, not annoying, and minimally intrusive), and the advertising (as little as possible). Obviously, you can tell I'm more of a public radio listener, and if I owned a commercial radio station I'd probably go bankrupt in a year. But I'd at least try to do things differently, as much as possible, and strive to capture the listener's imagination, create the conditions for solid brand loyalty, and figure out other ways to generate revenue from this committed audience besides continuously assaulting them with obnoxious DJ's, station promos and commercials.

"You may say I'm a dreamer/but I'm not the only one." As I said above, change is in the air, so my idealist vision may not be too far off. For instance, if a radio equivalent of TiVo were to come out, for home and auto, and listeners could easily skip all commercials and perhaps even DJ chatter, well, that would spell the end of commercial radio as it exists today, I should think. And it really doesn't seem to be a question of "if," but of "when." Radio will probably move to a subscription model: listeners will subscribe to a service, as they do now with cable, satellite radio, or TiVo, which will give them access to a wide variety of stations with many different formats and little or no commercial interruption. When this happens, the naming/branding of individual stations (formats) will likely become less important—they'll just be descriptively named to identify the format—while the naming/branding of the radio access providers will become extremely important, because competition among those players will be fierce.

Congress and the FCC will most likely determine how the structure will evolve, and whether radio as a medium remains in the hands of a few large corporations like Clear Channel and the satellite operators, or if it's opened-up to companies of every size broadcasting over the Internet that listeners can receive with high-bandwidth wireless IP receivers in their cars. What's clear is that as users taste the freedom to control their media, through the Internet or services like TiVo, they are increasingly unwilling to go back to the old way. It's what obnoxious brand consultants call a "paradigm shift," but you didn't hear that from me. Stay tuned.

If a radio station hired your company, what could you do for them?

As with a company or product naming project in any other industry, we would first do a thorough analysis of the names and messaging of all the competition to find out what's been done, what's been done to death, what areas are ripe for exploring, and the edge that everyone else is afraid to cross. We would probably advise against being restricted by call letters, or playing-off the type of music being played ("EZ 103" or "Lite 99" for easy listening, "KMTL" or "Power 104" for heavy metal, etc.). The goal would be to differentiate the station from the pack and create a brand that is bigger than the goods and services being offered—in this case broadcast radio and all related spin-offs.

We would get involved with the station to learn everything we could about its culture, format, history, listeners, disc jockeys, revenue, advertising, record-company alliances, partners, concert promotions, etc, to fully understand where they are coming from and where they need to go. We would begin developing the new brand positioning and new name options simultaneously—for us, the naming and the positioning go hand-in-hand, and each informs the development of the other. As we narrow the field of name candidates, we would present them to the client with contextual graphic support—ads, websites, audio clips, etc—to give them a life, to make them real, rather than just lists of names on a white page or screen. Obviously, with radio, the phonetic component of a name is more important than in most other areas, so we'd probably test a number of different audio treatments of potential names.

> "We would probably advise against playing off the music type you play (e.g., "EZ 103 for easy listening)"

Usually, when we find the right name for a new brand, it quickly becomes the perfect name, and all others fade away. The chosen name will be the final piece of a complex puzzle that is a brand—the brand positioning will dictate the kinds of names that will work, and the final name will reinforce the brand positioning. All of these steps are discussed in much greater detail in the Process section of our website (http://www.igorinternational.com). We also have a printable version of our process available as well.

Beyond naming, we help clients optimize their websites for search engines, which is really more of a messaging task than a technical task. We also write copy, design promotional activities, and effectively use the Internet to conduct grassroots viral marketing campaigns to spread the word about a company or product, as we recently did for BBC America's Golden Globe winning comedy show, *The Office*.

Create Evangelists

I love Mickey Mouse more than any woman I've ever known
—*Walt Disney*

Creating Customer Evangelists *is a book aimed at helping you build buzz, and Ben McConnell is its co-author. Buzz is, as I keep saying, the world's most effective form of advertising. And at the heart of buzz-making are your big fans, your "evangelists." Here's the how-to for radio.*

Ben, you and your co-author live in Chicago. In your estimation, what are some of the right things Chicago area stations are doing to create and sustain customer evangelism? What are they doing wrong?

> "Focus on a single cause year-round, not just during the holidays"

Several Chicago stations understand the value of building loyalty strategies, specifically via listener clubs. These self-selected groups are the most strategic opportunity for knowing who listeners are and understanding their motivations.

I've subscribed to a number of radio station loyalty clubs, and the writers of notes, emails and newsletters from many of the stations are invisible. They're nameless and faceless. They've been body-snatched by corporate aliens. Rule number one of customer evangelism for services: *People are loyal to people, not necessarily brands.*

The public radio station here, WBEZ, is becoming more sophisticated in its use of email marketing. Its newsletter is personal and relevant, written by Wendy Turner. She's fun and sprightly. She asks readers for their opinions. Wendy understands that personality-driven communications helps WBEZ cross the emotional chasm required to create listener evangelists. When WBEZ sends pledge pleas directly from on-air talent like "This American Life" host Ira Glass, the response rate is amazing.

WLUP-FM (the Loop) is very smart about marketing its "Loyal Looper" program. It promotes the heck out of it, and it appeals to the WLUP listener who literally has been listening to Classic Rock and the station for years, maybe a decade or more.

WKSU, a public radio station in Akron, Ohio, (and a client of ours) has created an online content product that focuses exclusively on folk music. Its popularity is growing rapidly; it has already exceeded year-end listener and registration goals and has a time-spent-listening average of 60+ minutes. Among the many reasons why it's growing quickly: It's innovative and unlike just about anything out there and it taps into the passion of a well-defined segment: folk music lovers.

What specific things can radio stations do to create evangelists?

Here are five:

1. If you don't have a loyal listener database, create one or sign up with a service. It's not an expense, it's an investment in your future success.

2. Know who your top 100 most loyal listeners are and treat them like the royalty they are. Royalty doesn't want discounts to local restaurants. Royalty wants access, which can mean many things—to on-air staff, station management, private audiences with advertisers and, of course, celebrities.

3. Focus on a customer-service culture. Create reward systems for employees based on attaining not just Arbitron results, but improved loyalty figures among listeners and advertisers. Extinguish any too-cool-for-school attitudes.
4. Focus on a single cause year-round, not just during the holidays. Put a stake in the ground and fight for it every week. The root of evangelism is creating emotional connections with your most affiliated customers; it doesn't have to appeal to everyone.
5. Create quarterly opportunities for advertisers to meet one another, and invite prospective advertisers to the parties. Watch how some existing, satisfied advertisers sell your station's services on your behalf.

What do you think is the biggest obstacle to evangelism for radio stations?

If you had asked this question before 1994, I would have said, "The inability to interact with listeners on a large scale."

With the Internet, radio stations can more effectively manage large-scale interaction. That said, the biggest obstacle today: To think less like a broadcaster and more like a relationship marketer.

> "Think less like a broadcaster and more like a relationship marketer"

Does it cost a lot of money to create evangelists?

Not at all. The strategies and tactics are what you make them. It does require:

- A belief in a theology, if you will. It's a leap of faith for many companies to think that customer evangelism can happen for them, but it won't happen unless you believe it can.

- A plan. Every radio station in the world should have this strategic objective: "To create more meaningful relationships with our customers (listeners and advertisers)." The strategies and tactics for that can be diverse, many of them free or very low-cost.
- Commitment and patience. It took Krispy Kreme decades to create the widespread customer evangelism it enjoys today.

What is the single most important reason radio stations should bother with evangelism tactics?

Two words: Satellite radio.

Would your listeners care if you went off the air tomorrow? A band of passionate believers can help prevent that scenario posed by satellite radio—wholesale usurpment of local content providers. What are you doing to become emotionally indispensable to your customers?

Look what's happening to network television, too: Viewership is eroding, especially this fall. The big question has been: Where did all the 18–34 year-old-men go? The networks blame Nielsen, but Nielsen says they're off surfing the Net, playing video games and being distracted by a confluence of new influences.

If radio doesn't prepare now, it will find itself more marginalized since the dawn of television.

Better Ads, Happier Listeners

He removes the greatest ornament of friendship, who takes away from it respect
—Cicero

Besides being the author of what I consider to be one of the best books on marketing and branding ever written (New Brand World), *Scott Bedbury is the former head of marketing for Nike and Starbucks. Scott knows marketing, he knows branding, and—like all listeners—he knows radio.*

> "A lot of Radio brands are getting hurt by their bad advertisers"

How would you characterize the quality of branding efforts by radio stations? Where are they going right? Where are they going wrong?

I applaud the efforts of any station that commits to branding itself in a consistent and thoughtful way. These days it's pretty dangerous not to try and set yourself apart from the fray. But there are some unique challenges for the radio industry that I find puzzling, and not because they're impossible but because no one seems to be addressing the core of the problem—the content of most radio advertising.

If the ads are obnoxious and inconsistent in terms of integrity, it's going to be pretty difficult to ask consumer to accept your station for anything other than a purveyor of unpredictable material.

The television networks aren't immune to this, but sometimes the biggest problem is the network itself. NBC tries to create "Family" viewers in its prime time, for example, but it uses that same period to advertise its own more adult rated programming that appears later, after the kids are in bed. It's not surprising that the first place the networks were able to brand was Saturday morning when the advertising had to follow certain rules. Nickelodeon took this to the bank. I trust Nick to respect my kids when they're watching. I can't say that for too many other networks or stations.

The radio industry suffers more in this regard because it's cheaper to produce radio advertising. Anyone with a couple thousand bucks can get on the air if they desire. Problem is, there don't appear to be many guidelines or thresholds as to the quality of the advertising that stations allow.

Now most stations, at this point, are probably wanting to shoot me for saying this, thinking that I just don't understand their business and the difficulty, especially in these times, of turning anyone away. I get that. I just can't accept that someone wants to build listeners and brand their station without looking at their entire package. Until recently, it was quite okay for most of them to ignore this problem. But technology will create greater problems for stations that continue to spray bad messages like bullets from an Uzi.

> "Clients should be accountable for the quality of their advertising"

If traditional radio doesn't address this, it will simply accelerate the consumer movement toward other options that technology will deliver, either through your own personal music player (think next generation iPod) or satellite radio. Honestly, who wants to listen to the latest screamer from a car dealer or a discount electronics outlet? Again, in the past there really wasn't much choice. You'd just hit the button to the next channel and probably find the same ad there a few minutes later. We'll soon have other choices.

But here's the thing that really bugs me: Clients should be accountable for the quality of their advertising. At some point they are accountable—it should have some effect on their sales—but most of them don't seem to care what it takes to get the sale. And I think a lot of radio "brands" are getting hurt by their bad advertisers.

But I think I have a solution. Hair-brained, maybe, but it's worth a try.

Imagine if a station allowed listeners to vote off a bad advertisement each week. They'd probably attack a slew of them, but the idea is that one ad each week would be singled out by the station as the one listeners really wanted yanked. They could vote on the station website (get ready for the traffic) where they could also vote on all the other ads, including the ones they liked. This would be valuable information for clients. They'd probably even pay for it.

If I were a radio advertiser I would like to know how I rate among all the other companies vying for the attention of listeners. If my ads really sucked I would want to know it.

But for this to work, the station would need to have the hair to boot the offensive ad off the air and force the client to produce something else. It would also require a dialogue between station sales and the advertiser if it looked like a client's ad was headed for the graveyard of bad ads.

"Yeah, Bill? This is Jeff at the KUBE. Your ad is really taking it in the shorts this week. You're 35% below the next lowest scoring ad. From the looks of it, you're probably damaging your brand. I realize it's only Thursday but it looks like we're going to have to hook it. As you know, every Monday morning we'll tell our listeners which ad we booted, so I thought you'd like to know ahead of time. Maybe you'd like to make a response. But I've got an idea: Why don't we do a promotion where we ask them to send you advertising ideas via email? It would be free. You know everyone thinks they're an advertising expert. Let's put 'em to work for you."

> "Respect the intelligence and time of listeners"

It could be the greatest radio promotion of all time. All the pent up frustration toward bad advertising unleashed. It would make national news. But it would require a station with a real backbone. A station that truly cared about its brand.

Here's the other reason for doing something like this. If the ad is bad it is probably forcing listeners away from the station every time it airs. I seldom tolerate more than a few seconds of stupid advertising before switching stations. Maybe it's just me, but I think there are a lot others out there like me that want to spend their time in better ways.

What is the difference between crafting a "positioning line" and developing a "brand"? Are the lines radio stations use even "positioning lines" or are they simply "slogans"?

You can create a positioning line in an hour. A brand takes years to develop. Brand development is a process in which everyone in the organization contributes in some way. Everyone supports the brand or they don't. Smart companies are taking brand development much more seriously these days. It informs how they develop products, how they advertise, what kind of people they hire. As a process it goes far beyond marketing. Some of the most important brand development decisions at Starbucks, while I was there, had to do with Human Resources.

What is the most important thing a radio station needs to do in order to develop a powerful, compelling brand?

Respect the intelligence and time of listeners.

Can you build a brand without spending a lot of money in marketing dollars?

Sure. If your message is impactful you can spend much, much less. At Nike we outperformed brands that spent five to ten times what we did in media, in terms of advertising recall and favorability. Our ROI was incredible. Some spots ran only a few times. We also did a lot of grass roots marketing, something we did at Starbucks when I was there.

> "If your product is not good you will need to spend a great deal of money to 'churn' consumers in and out of your brand"

If your product or service is not good you will need to spend a great deal of money to "churn" consumers in and out of your brand. Telephone companies like AT&T do this. They spend billions trying to get back customers they lose every year because their service is no different than anyone else, or they just have poor service.

When I arrived at Starbucks in 1995 the entire marketing budget was $3 million. And half of that was department overhead. In the years that followed, arguably the most intense growth period in the history of the company, we never spent much on advertising because we knew that our first priority was to respect our customers and create a great experience for them in the stores. There would be no point in spending lots of money driving customers into a coffee house that was disrespectful, uncomfortable or inconsistent. The coffee experience was our "content." We took it very seriously. It was the core of our marketing program.

There is a great analogy in here for any radio station. What kind of experience are you providing? Not just the content you control but everything the consumer hears. If the experience is great, you won't have to spend much in terms of marketing.

Inspiring Passion

I love you Hilary Duff!
—*Lindsay Lohan*

Charlie and the Chocolate Factory *is a huge hit at the movies, and that legendary chocolatier Willy Wonka has some great advice for radio marketers.*

Mr. Wonka, candy fans go nuts for your products. How can radio stations inspire that kind of passion?

> "My job isn't to defend what I have, it's to create exciting products for the future"

In my estimation, radio stations tend toward one and only one flavor, vanilla. I'm a businessman and have the same financial pressures every radio station has. But I also understand that innovation and risk-taking are what set me apart from my competitors. My job isn't to defend what I have, it's to create exciting products for the future. Your audience will get excited when you do things designed to excite them.

I created a chewing gum that never loses its flavor and another that tastes like a four-course meal. I made an ice cream that doesn't melt—even in the sun. These are things my competition can't even imagine, and it's because they don't try. Yet they're exactly what inspire so much passion for my products.

Radio stations need to understand that risk and reward always have and always will go hand in hand. And innovation is inseparable from risk-taking.

I ask "why can't I…?" So should you.

Why is your "Golden Ticket" contest so unlike a typical radio station contest?

Radio station contests are usually about cash prizes. I know that's partially because that's what people tell you they want to win. It's also because cash works—but not always. In fact, does it work even half the time? And does it work better than anything else? Or just better than anything else you bother to imagine?

What people need to solve their everyday problems and what fires their imagination and fuels their dreams are two very different things. Yet it's the latter we live for. The former just helps us subsist. When you can fulfill a fantasy you haven't just awarded a prize, you've transformed a life.

> "When you fulfill a fantasy you haven't just awarded a prize, you've transformed a life"

My prize was all the chocolate you can eat and what I called "an extra prize beyond your wildest imagination." Charlie was tempted to sell his winning ticket to the highest bidder, but as his grandfather told him, "there's plenty of money out there, but this ticket, there are only five in the world. Only a dummy would give this up for something as common as money."

There seems to be a lot of mystery about your company. Can that air of mystery work in radio, too?

Nobody had been inside my chocolate factory for a generation. Part of the thrill of the contest was peering inside those mysterious walls.

Sometimes what listeners don't know can create intrigue and curiosity. When I listen to Howard Stern (don't tell the kids I listen to Stern!), it sometimes takes half an hour to know who he's talking to. I used to think that was "bad radio." But now I see it differently.

Figuring out who Howard is talking to is like a puzzle, a mystery. It engages the minds of listeners to struggle through this puzzle and that keeps their ears pinned to the radio. It's what they don't know, sometimes, that keeps them listening.

Mystery can sell radio as well as it can sell chocolate.

The Future of Radio

A great empire, like a great cake, is most easily diminished at the edges
—Benjamin Franklin

I created this book because I love radio. Because I think there's something deeply resonant and important and wonderful about what radio provides its audience. But my greatest fear is that our industry will be a victim of its own success, its own hardened arteries and calcified ligaments.

Now is the moment in the history of radio when we need to think smart about our industry's future. And that starts with an obvious but completely misunderstood question:

What business are we in?

Buggy manufacturers disappeared from the face of the Earth because they thought they were in the buggy business, not the transportation business. With the swarm of up-and-coming technologies rocking the headlines and our world daily, with satellite, iPods, podcasting, streaming broadband, etc., all threatening to slice off a portion of our listening audience, it's fair to ask what turf we should be defending. It's fair to ask, as an industry, where is radio going?

Fundamentally, we all need to understand one thing loud and clear:

> "What business are we in?"

We are not in the "radio" business.

If your station is all about music, sports-talk, talk, or personality you're in the *audio entertainment business*.

Not the "local" audio entertainment business. Not the "mobile" audio entertainment business. Not the "free" audio entertainment business. These are all unnecessarily small niches. Historically, we have had the market almost to ourselves, especially whenever the listener was away from home. Our only music competition came from the occasional CD or cassette or vinyl disc. And personality radio had no competition at all. That's all changing folks. And it means we have to change, too.

Being in the audio entertainment business does not just mean that all the evolving, competing technologies are threats; it also means you must exploit these very same technologies for your own benefit.

When the Internet blasted onto the scene, Microsoft didn't lay there and say "so what, we're a software company" (as we're saying: "So what, we're radio, we're free, we're local"). Oh no. They said "we're in the business of helping computer users get their work done, and this Internet thing changes our business strategy completely. We must exploit it."

There's no advantage for audio entertainment to be local. Especially for music-intensive stations where their programming is a commodity easily reproducible in a multitude of forms. Even for personalities, it will take a great local talent to stem attrition to nationally-based greater talents. There's

> "Opie and Anthony and Howard Stern can move to Satellite if they want, but it is only because of Radio's vast reach that any listener will care"

a reason, for example, that no local TV affiliate anywhere in the country has any significant local entertainment programming: The national stuff is clearly and unambiguously better. Talent is and always has been scarce. And it will invariably gravitate to a broad *national* audience.

If "local" isn't an advantage to the audio entertainment business, if "free" isn't an advantage in a world where folks gladly pay for content that's valuable (i.e., Cable TV or bottled water), if "mobile" is fast-becoming a technological free-for-all, then what inherent advantages do we have?

We have *universality*.

For now, at least, everyone listens. That means we have vast distribution. Much like a TV network, we're already in every home—and we're also in every car and on every hip. Opie and Anthony and Howard Stern can move to Satellite if they want, but it is only because of radio's vast reach that any listener will care. Subtract the reach and you subtract the personality.

And once you have reach—universality—I believe you should *add* the personality. And not the home-grown organic Joe Blow variety that generally takes two years to gel—if it ever gels at all. I mean the Grade A, Hollywood Comedy Star variety. And do it nationally. Because, once again, there is no "local" advantage in the audio entertainment business. Does this involve terrific risk? You bet. National radio will be a lot more like network TV. And as any TV executive knows, no risk equals no rewards.

I say "free" is not an advantage because ultimately we're foolish if we don't offer premium content and charge listeners a fee to hear it. Some morning shows are doing this now. O'Reilly and Limbaugh do this online. It's inevitable, folks. We should give away most of our product to listeners for free so we can sell them the rest.

But what about those stations which aren't music-based?

If your station is all about news, sports, traffic, or weather then you're in the *audio information* business.

Being "local" is definitely an advantage when you're in this business because information, like politics, is primarily local. This is why, for example, local TV stations are overgrown with locally-based news programming.

In the audio information business, technology is conspiring to rid you of your proprietary advantages. The Internet has weather and traffic—and has a picture of it. Satellite radio is making deals to localize its information services. You must climb aboard the techno-train and exploit any and all available technological routes to maintaining your position. More than any other form of radio, audio information must embrace technology. And radio companies with a big stake in audio information stations should be buying up and making deals with relevant technology vendors.

> "When was the last time you heard of a radio broadcasting company buying an Internet firm in the business of audio entertainment or information?"

Our vast distribution is an advantage—only for a while. After all, the Internet's distribution is likewise vast. What the Internet lacks, however, is *warmth*. Ask any NPR listener and they'll tell you that they love NPR not only because of the information but also because of the personalities providing it. *I believe information personalities are as important as information content.* Do you have personalities or do you have newsreaders?

The main point you should understand clearly is this one: *Whether or not you're in the audio entertainment or information business, you are competing against any other player in any medium which can likewise be termed audio entertainment or information.* That means that other radio stations alone are not your competition. Satellite radio alone certainly isn't your competition. The cell phone is also your competition. So is iTunes. So is any and every source for wireless audio entertainment and information.

Right now our industry is bigger than any of those competitors. And that means we should be investing heavily in alternative windows to reach the ears of our audience. When was the last time you heard of a radio broadcasting company buying an Internet firm in the business of audio entertainment or information? Instead, Yahoo or Google buy them. AOL buys them. Sprint and Verizon do deals with them.

We are blind to the future and it's right in front of our eyes.

The great white hope of the radio industry is assumed to be digital radio. But this is the wrong road. And it will distract us from the real opportunities for years to come.

So-called "HD radio" is unlikely to lead us to the Promised Land. It has at least eight major marketing problems (to say nothing of its technical issues):

Problem 1: How do you sell a radio?

HD radio requires that consumers buy new hardware. New radios. Not since the dawn of FM have we had to sell radios—and that was a specific moment in time when a dramatically improved fidelity and the hottest music in the land conspired to change the way America listened to the radio. The difference between FM and HD is not nearly so great to the average ear as the difference

> "How do you sell a radio?"

between AM and FM was. And there is no hot music waiting to be discovered by brand new radios.

Without a clear audience need we simply don't know how to move hardware. We're content providers. And we own the pipes. We deal with a distribution channel that's already in every car, workplace, and home—it's universal. We don't know—and haven't needed to know—how to sell radios. And if you think selling these radios is as simple as "getting behind the effort," talking it up, and handing out free samples, please stop taking the brown acid now.

If people are to buy these radios it will be for one of two reasons:

a. Either the radios will piggyback on something they buy for other reasons (i.e., I buy a new car and HD radio comes standard) or…

b. They will want the content that is available exclusively on HD radio (and nowhere else), an unlikely scenario in today's ultra-connected world

Problem 2: HD radio is Selfish

HD radio presumably solves an industry problem, namely how to keep up with technology, expand our offerings to advertisers, and compete more effectively with Satellite radio.

But what audience problem does it solve uniquely?

Here is a slate of competing answers, all of them flawed:

a. **More choice?**

Listeners may *say* they want more choice, but in the real world of consumer behavior abundant choice is a distraction. Too many choices are confusing, and when faced with confusion most customers do what's easiest and most comfortable: They choose the tried and true.

It's a phenomenon author Seth Godin calls a "scarcity shortage." J. Walker Smith, president of Yankelovich Partners, has called it a "claustrophobia of abundance." In the real world, consumers prefer fewer options, not more. Small wonder there's a new breed of boutique out there where the selection of product is limited to the discriminating taste of owner. In a sense, it's like a listener trusting her radio experience to the judgment of her favorite DJ who picks the records himself.

It is critical that the radio industry understand this concept: The vast majority of listeners already have enough choice—all our research and ratings experience tells us that. We assume that listeners subscribe to Satellite radio because of the enormous choice. I think this assumption is wrong. The statistics indicate that satellite listeners tune in only slightly more channels than radio listeners tune in stations. Having access to 100 channels is not an attraction if I don't use them.

> "The vast majority of listeners already have enough choice"

But choice isn't just about using more channels or stations, it's about having more unique ones to choose from, you might argue. True enough. But would it surprise you if you discovered that the most popular satellite radio channels are clones of the most popular radio stations?

Even in a world of more than 100 channels, the hits are the hits.

Granted, if there's an obscure type of music stream that exists only on satellite, that's a perfectly attractive reason to subscribe. But now you're into the world of niche programming, and there are many more niches than multicasting will allow. And radio will hardly be unique in servicing those niches. Remember, niches are, by definition, small.

Wired editor Chris Anderson tackled this topic in a seminal piece, *The Long Tail*. Chris used the example of the music download service Rhapsody: "Not only is every one of Rhapsody's top 100,000 tracks streamed at least once each month, the same is true for its top 200,000, top 300,000, and top 400,000. As fast as Rhapsody adds tracks to its library, those songs find an audience, even if it's just a few people a month, somewhere in the country. This," Chris wrote, "is the Long Tail."

Niches can be attractive—but maybe to "just a few people a month, somewhere in the country." For a broadcast medium built on an advertiser-supported model, sliding down the long tail will drop us deep into shark-infested waters.

We have a deep industry bias leading to the mistaken conclusion that "more variety is what listeners want" despite abundant evidence to the contrary.

As of this writing, the newest and hottest Apple iPods contain *less* capacity, *not* more. That is, they allow *less variety*. The iPod Nano can fit only 1,000 songs while the new Motorola iPod cell phone fits only about 100, which surveys show is exactly the number of songs the average person keeps on his or her mp3 player *regardless* of its capacity.

Yes, that's right. The average person puts fewer songs on his or her mp3 player than on any station you can possibly listen to. This should surprise no one who understands the dynamics of "hits," so why is radio chasing the "long tail"?

b. **More versions of your core format?**

This is the same as "more choice." Besides, dozens of format skews are already available on Satellite radio or via streaming

> "The newest and hottest iPods allow *less* variety, not *more*"

broadband—even to your cell phone, the ultimate "portable radio." And by the time HD radio is widely available these alternatives to conventional radio will likely be flourishing.

Besides, when you launch a new HD "Classic Country" channel beside your existing "Mainstream Country" channel, where do you think your listeners to the new channel are most likely to come from? That's right, your old channel. The result: More cannibalization, lower ratings. We will be eating our young.

c. **"It's Digital!" i.e., better audio quality?**

To the industry (and any consumer who does their homework) HD radio is positioned as "Pure Digital. Clear Radio." The pitch, in other words, is that this is technologically better radio. Where's the evidence that this fine distinction in audio quality is a meaningful benefit, that "bad audio" is one of radio's audience problems? Most people don't have a problem with the audio quality of their radios. The vast majority of your audience is not comprised of audiophiles. In fact, your listeners are less likely to be discriminating musicologists and more likely to be tone deaf.

> "What's the right price for an anvil? Well, if you don't need an anvil then zero is too much"

As Marketing guru Seth Godin wrote me on this topic: "Yikes, [audio quality as a benefit for HD radio] is such a hard sell. I just spent thousands of dollars LOWERING the quality of my stereo at home by switching my CD's to MP3s and buying a Sonos player. The iPod vastly outsells turntables because people don't want quality, they want control."

And where's the "control" with HD radio?

In the hands of broadcasters, that's where, not listeners.

d. **No commercials?**

Commercials: Now there's a listener problem. But HD radio won't help us there.

e. **Data channels for services such as traffic, news, and weather?**

Those services are already on the market and will be commonplace by the time HD radio is widely available. Besides, how many traffic, news, and weather channels does one market need?

f. **But Radio is "Free"!**

Indeed it is. But time and again consumers prove they're willing to pay for what they value. After all, water is free yet people buy the expensive bottled stuff every day—more than $8.3 billion in 2003 alone. And that's in spite of evidence proving that what comes from a bottle is no better than what flows from the tap.

Besides, just try telling folks HD radio is free in the same breath you tell them a new radio costs $200 dollars.

As Godin says, what's the right price for an anvil? Well, if you don't need an anvil then zero is too much.

So again, we may perceive that HD radio solves our industry's problem? But what listener problem does it solve?

Problem 3: What's the Simple, Clear Benefit?

People buy what benefits them—in fact, in a real sense people buy the benefit, not the product. They buy products that satisfy wants better than other products do and they buy stuff which satisfies those wants uniquely. If I as a consumer don't have a problem with radio I don't need HD radio.

This, in fact, is why there aren't already millions more subscribers to Sirius or XM: Most people don't have a radio problem. And no problem requires no solution. And no solution means no HD radio. As Godin writes in his recent book *All Marketers are Liars*, "If you're walking around believing you don't have a radio problem, then the greatest radio solution in the world isn't going to show up on your radar. It's invisible."

So what listener problem does HD radio solve? And what is the single sentence which sums up the way in which HD radio solves that problem in simple, clear, compelling terms?

Problem 4: Who is the Target Audience?

Who are the best first customers for HD radio? They are the tech-heads, the innovators, the gadget freaks, the trendsetters.

They are, in other words, the very same folks who took the dive with Satellite radio and are now under contract with Sirius or XM. They are the same folks who pay Launch.com or MSN to stream their "radio" online. Either way, they will be disinclined to cancel just so they can buy yet another radio—unless our content is utterly unique.

> "Becoming more like Satellite Radio means becoming redundant to Satellite Radio"

Problem 5: Fighting Satellite radio on their terms is a losing proposition

There's a logic that says because Satellite is nipping at radio's heels we must take action to be more like Satellite radio. This logic ignores the fact that becoming "more like" Satellite means becoming redundant to Satellite radio. And being redundant to

Satellite radio means being an also-ran with no significant positioning advantage and doing it much, much later than Satellite. Yahoo isn't good at being eBay, Buy.com isn't good at being Amazon, and radio won't be good at being Satellite.

Problem 6: The Product is Different in Every Market

Satellite radio is a coherent brand, the same nationwide. HD radio will be different in every market, depending on the programming decisions made in that market.

That means that Satellite radio is selling one product but HD radio is selling as many products as we have markets.

How do you sell a technology that is different everywhere you sell it? Answer: You can't. You can only sell the content on it. That adds a sizable quantity of confusion to the mix, and the more confusing the product the less consumers will pay attention to it. And the less attention they pay to it, the less apt they are to adopt it.

Problem 7: The Technology "Cart" is before the Content "Horse"

We should know what we're going to multicast on this technology before we set out to market it. Consumers aren't buying radios, after all, they're buying what we put on them. When was the last time you bought a ticket to a show without knowing who was playing?

> "Radio doesn't live in a vacuum"

Why are we as an industry inviting our audience to a "blind date" with our future?

It is the content in the technology—what we put on the radios—not the presumed "gee-whiz" factor that have always made radio stations hits. And make no mistake: That content must be special and magnetic and unique. Splintered versions of our existing formats will not be sufficient.

Radio's future will demand star talent and will, I predict, be driven by non-music content—talk and entertainment.

Problem 8: Radio doesn't live in a Vacuum

HD radio is not simply the multiplication of choice to radio listeners. It's the creation of an all new set of "pipes" to deliver content. And these pipes—like the ones currently delivering radio—are owned and licensed by radio broadcasters. In other words, HD radio is a means for the radio industry to perpetuate control over the audience's listening choices in an era when there are many routes to listener ears—not just the ones owned by radio.

But if you're really in touch with the customer, the listener, why is there a need to create a new set of "pipes" when "pipes" are proliferating at a rapid pace already and there are plenty of "pipes" out there on the market right now?

The bad answer is simple: Because radio doesn't own these "pipes."

You can't put the technology genie back in the bottle, folks. Pandora's box has been opened and there's no turning back. The radio industry no longer has exclusive control of audio entertainment and information and it never will again. Inventing a new distribution channel which is redundant to existing channels is like shouting into the wind. It's unlikely the audience will hear.

Thus, HD radio is already competitively outfoxed, before it even gets out of the gate

The phenomenon of Podcasting will allow listeners to get both music and non-music content while bypassing radio of any kind. I know it's a licensing nightmare right now and the technology required to Podcast is not yet mainstream, but both of these issues will be resolved within a matter of months, well ahead of HD radio's rollout.

Meanwhile, my sister doesn't listen to radio. Her office, her co-workers, and the hundreds of listening quarter-hours they represent every week belong to streaming audio, which will only grow as high-speed connections become ever more ubiquitous.

> "Talent is the key to radio's future"

And that only scratches the surface. High-speed Internet connections will soon be in cars, and high-speed Internet cell phones are on the horizon. But, unlike radios, cell phones are commonly upgraded every two years.

Within the next twenty-four months WiFi will be widely available and FREE to all in cities like Austin, Portland, Philadelphia, New York City, San Francisco and others. And that's not all. Reuters reports that "Slightly more than 100 US cities…are setting up wireless networks now…[and] close to 1,000 local governments worldwide have plans in the works."

Free wireless audio access?

There's another name for that:

"Radio."

So what should we do?

That's the most common question I receive in light of the blinding changes our industry is facing.

And the answer is as simple as it is obvious:

We must leverage our financial prowess and our vast distribution with the most compelling and unique content available. It is patently ludicrous that radio should lose name brand talent to Satellite when radio makes money hand over fist while Satellite has yet to see any color ink but red.

Talent—off-air and especially on—is the key to radio's future.

Radio must be more entertainment and personality-oriented. Radio must take more chances. Radio must plan fall seasons in the same way that TV does. Radio must think in terms of programs, not just programming. Radio must re-think the four and five hour day-part. Radio must spend the kind of money on top-notch air talent that only we can afford and justify. Radio must give listeners a reason to stay tuned, no matter where they listen or what gadget they listen on, no matter what's on their iPod or their cell phone, no matter what's streaming on AOL. We must not cling to HD radio like a castaway to a piece of floating wreckage. Instead we must embrace all manner of technologies that help us fulfill a core mission: To connect the ears of the mass audience to the best, most distinctive, original, and attractive programming available.

To be a destination worth tuning in and worth working in.

We must build brands, not mix songs. We must add value, not cut costs. In the final analysis we're in the *content* business, the *remarkable* content business not the *cheap* content business.

If we don't come to terms with this fundamental notion, we'll see our audience slip away to more compelling, original alternatives. We'll be left with the technological laggards who lack access to better content: An older, poorer, less educated audience—exactly the people advertisers are least interested in reaching.

Recall Doug Hall's advice: "Think of your station more like a movie studio than a movie theater. Rather than a projector, you are the creator."

Our mandate is to create distinctive content that thrills and delights and touches and comforts and entertains and informs, and to do it wherever audio exists. It's a powerful offense and a mighty defense.

If we're not in this business to make a difference in the world, then stop the world now and let's all get off.

The fresh air is where the magic lives.

978-0-595-37658-2
0-595-37658-4

Printed in the United States
40876LVS00001B/46-144